EQUATIONS OF MIXED TYPE

EQUATIONS OF THE MIXED TYPE

A. V. BITSADZE

Translated by

P. ZADOR

Translation edited by

I. N. SNEDDON

Simson Professor of Mathematics in the University of Glasgow

A Pergamon Press Book

THE MACMILLAN COMPANY

NEW YORK

1964

THE MACMILLAN COMPANY
60 Fifth Avenue
New York 11, N.Y.

This book is distributed by
THE MACMILLAN COMPANY
pursuant to a special arrangement with
PERGAMON PRESS LIMITED
Oxford, England

Library of Congress Catalog Card No. 61-11158

This is a translation of the original Russian *Uravneniya smeshannogo tipa* published by the Academy of Sciences of the U.S.S.R., Moscow, 1959

Made in Great Britain

CONTENTS

To my dear teacher,

Mikhail Alekseevich Lavrent'ev

FOREWORD

The theory of equations of mixed type originated in the fundamental researches of the Italian mathematician Francesco Tricomi which were published in the twenties of this century.

Owing to the importance of its applications, the discussion of problems concerned with equations of mixed type has become in the last ten years one of the central problems in the theory of partial differential equations.

The present work is not meant to be a summary of all results in this field especially since the number of results increases with great speed; nevertheless the reader of this monograph will obtain an idea of the present state of the theory of equations of mixed type.

This book was developed from a series of lectures dealing with certain fundamental questions in the theory of equations of mixed type which the Author delivered in scientific establishments in the Chinese People's Republic at the end of 1957 and the beginning of 1958.

The Author wishes to express his thanks to M. A. Lavrent'ev, I. N. Vekua, L. D. Kudryavtsev and to the collective of research workers in the department of mathematical physics of the Mathematical Institute of the Academy of Sciences of China, headed by Professor U Chin-Mo, who made extremely valuable suggestions during the course of the preparation of this monograph. The Author is also indebted to M. M. Smirnov, S. A. Tersenov and A. Z. Ryvkin who after reading the manuscript made a series of useful comments.

INTRODUCTION

In the theory of partial differential equations of the second order

$$\sum_{i,k=1}^{n} A_{ik} \frac{\partial^2 u}{\partial x_i \, \partial x_k} + \sum_{i=1}^{n} B_i \frac{\partial u}{\partial x_i} + Cu = f, \tag{1}$$

where A_{ik}, B_i, C and f are given real functions defined in a certain domain D of the space of real variables $x_1, x_2, \ldots, x_n$, the so-called characteristic quadratic form

$$k(\eta_1, \eta_2, \ldots, \eta_n) = \sum_{i,k=1}^{n} A_{ik} \eta_i \eta_k \tag{2}$$

plays an important part.

If at the point P $(x_1, x_2, \ldots, x_n)$ of domain D not all coefficients A_{ik} are zero, i.e. if the order of equation (1) is not degenerate, then, by means of the affine transformation

$$\eta_i = \sum_{k=1}^{n} l_{ik} \xi_k, \quad i = 1, 2, \ldots, n$$

the quadratic form (2) can be reduced at the point P to the canonical form

$$k = \sum_{i=1}^{n} a_i \xi_i^2,$$

with the coefficients a_i assuming the values $+1$, -1 or 0. The number of negative coefficients (the so-called index of inertia) and the number of vanishing coefficients (the so-called defect of the form) are affine invariants.

Correspondingly, equations of form (1) can be classified into types. It is said that equation (1) at point P is: (a) of the elliptic type if the a_i are either all negative or all positive; (b) of the hyperbolic type if one of the a_i is negative and the rest are positive (or conversely); (c) of the ultrahyperbolic type (for $n > 3$) if m coefficients a_i ($m \neq 0, 1, n-1, n$) are positive and the remaining $n - m$ are negative; (d) of the parabolic type if at least one of a_i is zero.

Equation (1) is elliptic, hyperbolic, or parabolic in domain D if it is respectively elliptic, hyperbolic or parabolic at each point of that domain.

A real non-singular transformation of the independent variables

$$x_i = \sum_{k=1}^{n} g_{ik} y_k,$$

can always be found to transform the main differential part of equation (1) at each point of domain D to the so-called canonical form

$$\sum_{i=1}^{n} a_i \frac{\partial^2 u}{\partial y_i^2}.$$

If the number of independent variables $n > 2$, then, generally speaking, it is impossible to derive such a canonical form by means of one and the same transformation even in an arbitrarily small domain of the independent variables. However, for $n = 2$, such a transformation exists under sufficiently general assumptions about the coefficients A_{ik} and the domain D in which equation (1) is considered.

For equation (1) classical problems such as, for instance, the problems of Dirichlet and Neumann (in the elliptic case), the problems of Cauchy and Goursat (in the hyperbolic case), the problem of Cauchy (in the parabolic case) and also some of the so-called mixed problems (both in the elliptic, and the hyperbolic and parabolic cases) have been studied in detail.

In applications we not unfrequently encounter the situation when in different parts of the domain, equation (1) belongs to different types. We then say that equation (1) is an equation of mixed type.

It is one of the most important problems of mathematical physics to study the properties of solutions of equations of mixed type. The first fundamental results in this direction were obtained by the Italian mathematician Francesco Tricomi in the nineteen twenties [1—4].

At present there are several mathematical works both by Soviet and foreign authors, in which a series of problems concerning linear partial differential equations of the second order in two variables and possessing the property that the type of the equation changes either on the boundary of or inside the considered domain, have been stated and investigated.

Below we shall treat several basic results from this comparatively new branch of the modern theory of partial differential equations.

CHAPTER 1

GENERAL REMARKS ON LINEAR PARTIAL DIFFERENTIAL EQUATIONS OF MIXED TYPE

§ 1. Equation of the Second Order with Two Independent Variables

Suppose that the coefficients A, B, and C of the linear differential equation of second order

$$A \frac{\partial^2 u}{\partial x^2} + 2B \frac{\partial^2 u}{\partial x \partial y} + C \frac{\partial^2 u}{\partial y^2} + A_1 \frac{\partial u}{\partial x} + B_1 \frac{\partial u}{\partial y} + C_1 u = f, \tag{1.1}$$

are not zero simultaneously at any point of domain D, in which this equation is considered.

The curves $\varphi(x, y) = \text{const}$ where $\varphi(x, y)$ is the solution of the equation,

$$A\varphi_x^2 + 2B\varphi_x \varphi_y + C\varphi_y^2 = 0,$$

are called the characteristic curves, and the direction $\mathrm{d}y/\mathrm{d}x = \lambda(x, y)$, defined by the equation

$$A\mathrm{d}y^2 - 2B\mathrm{d}y\,\mathrm{d}x + C\mathrm{d}x^2 = 0 \tag{1.2}$$

the characteristic direction of equation (1.1).

Depending on whether in the considered domain the discriminant $AC - B^2$ of the quadratic form (1.2) is larger than, smaller than or equal to zero, the equation (1.1) is said to be elliptic, hyperbolic or parabolic.

It follows from the differential equation (1.2) of characteristic

$$\frac{\mathrm{d}y}{\mathrm{d}x} = \frac{B}{A} \pm \frac{1}{A}\sqrt{B^2 - AC} \tag{1.3}$$

that at points where equation (1.1) is elliptic there are no real characteristic directions, whereas at each point where it is hyperbolic there are two, and at each point where it is parabolic there is one.

If equation (1.1) is of the same type throughout the considered domain then, as is known, we can find a non-singular real transformation of the independent variables $x = x(\xi, \eta)$, $y = y(\xi, \eta)$ which will transform the terms involving second order partial derivatives of this equation in the domain D to the following canonical forms:

$$\frac{\partial^2 u}{\partial \xi^2} + \frac{\partial^2 u}{\partial \eta^2} \quad \text{(in the elliptic case}$$

$$\frac{\partial^2 u}{\partial \xi^2} - \frac{\partial^2 u}{\partial \eta^2} \quad \text{(in the hyperbolic case)}$$

$$\frac{\partial^2 u}{\partial \xi^2} \quad \text{(in the parabolic case).}$$

It can happen that in different parts of domain D equation (1.1) belongs to different types.

As was already pointed out, equation (1.1) is parabolic whenever

$$\Delta(x, y) = AC - B^2 = 0. \tag{1.4}$$

Assume that coefficients A, B, and C of equation (1.1) are smooth functions and that the points which satisfy equation (1.4) in domain D, form a simple (smooth) curve σ.

If in the considered domain D, containing the curve σ, equation (1.1) is everywhere elliptic or everywhere hyperbolic except on σ, then we say that the domain D is an elliptic domain or an hyperbolic domain, respectively, with a parabolic degeneracy along σ. If, however, σ divides domain D into two parts such that in one of them equation (1.1) is elliptic and in the other it is hyperbolic then we say that equation (1.1) is of mixed type in the domain D. We then sometimes say that a D is a mixed domain.

Denote by α the smallest angle between the tangent to the curve σ at point P and the characteristic direction (1.3) at the same point. We shall see later that the solution of equation (1.1) behaves along the line of degeneracy σ in essentially different ways according as $a = 0$ or $a \neq 0$.

As we know, it is considerably simpler to study elliptic, hyperbolic or parabolic equations of the second order with two independent variables after they have been reduced to canonical forms. Naturally, the reduction to the canonical form of equation (1.1) by means of a non-singular transformation of the independent variables in the presence of a line of degeneracy in the considered domain is a problem of first-rate importance. Extremely valuable researches in this direction are reported in the works (1,2) of the Italian mathematician Cibrario.

§ 2. The Theory of Cibrario

Since the type of an equation of the second order depends solely on the terms involving the partial derivatives of second order and the problem is precisely that of reducing to the simplest form a sum of such terms (the dominant part of the equation), it is natural to consider instead of (1.1) the more general equation

$$A \frac{\partial^2 u}{\partial x^2} + 2B \frac{\partial^2 u}{\partial x \, \partial y} + C \frac{\partial^2 u}{\partial y^2} = E\left(x, y, u, \frac{\partial u}{\partial x}, \frac{\partial u}{\partial y}\right), \qquad (1.5)$$

where A, B and C are given functions depending only on variables x and y.

Suppose that A, B, and C are real analytic functions in some domain D of the real variables x and y, and that the set of points in D, at which parabolic degeneracy occurs is a simple (analytic) curve σ.

Suppose that all partial derivatives of the function $\Delta(x, y)$ up to and including the $(n-1)$th vanish along the curve σ, and that at least one of the nth derivatives is non-vanishing. In view of the fact that the derivatives $\partial^n \Delta/\partial x^n$ and $\partial^n \Delta/\partial y^n$ are not simultaneously zero, along the curve σ the function $\Delta(x, y)$ can be represented in some neighbourhood of that curve in the form

$$\Delta = H^n(x, y)\, G(x, y), \qquad (1.6)$$

where $H(x, y) = 0$ is the equation of σ, and function $G(x, y) \neq \neq 0$. Noting at the same time the fact that the derivatives $\partial H/\partial x$ and $\partial H/\partial y$ cannot simultaneously vanish on σ, it is possible to find a subdomain $\delta \subset D$, containing the curve σ in which the representation (1.6) is valid and also $\partial H/\partial x$ and $\partial H/\partial y$ are not simultaneously zero at any point of domain δ.

On applying the transformation of the independent variables

$$\xi = \xi(x, y), \quad \eta = \eta(x, y) \tag{1.7}$$

the dominant part of equation (1.5) assumes the form

$$A \frac{\partial^2 u}{\partial x^2} + 2B \frac{\partial^2 u}{\partial x \, \partial y} + C \frac{\partial^2 u}{\partial y^2} = (A\xi_x^2 + 2B\xi_x \xi_y + C\xi_y^2) \frac{\partial^2 u}{\partial \xi^2} +$$

$$+ (A\xi_x \eta_x + B\xi_x \eta_y + B\xi_y \eta_x + C\xi_y \eta_y) \frac{\partial^2 u}{\partial \xi \, \partial \eta} +$$

$$+ (A\eta_x^2 + 2B\eta_x \eta_y + C\eta_y^2) \frac{\partial^2 u}{\partial \eta^2}.$$

We shall try to choose ξ and η in such a way as to make the transformation (1.7) non-singular and, in addition subject to the conditions

$$A\xi_x \eta_x + B\xi_x \eta_y + B\xi_y \eta_x + C\xi_y \eta_y = 0, \tag{1.8}$$

$$A\eta_x^2 + 2B\eta_x \eta_y + C\eta_y^2 \neq 0. \tag{1.9}$$

There are two possibilities for points on the curve σ either

$$AH_x^2 + 2BH_x H_y + CH_y^2 \neq 0 \tag{1.10}$$

or

$$AH_x^2 + 2BH_x H_y + CH_y^2 = 0. \tag{1.11}$$

The inequality (1.10) means that the direction of the characteristic of equation (1.5) is not the same for points on the curve σ as the direction of the tangent to this curve, whereas identity (1.11) implies that these directions coincide.

Suppose first that the inequality (1.10) holds everywhere along σ. In this case as the variable $\eta(x, y)$ we take the function

$$\eta(x, y) = H(x, y),$$

and choose the variable $\xi(x, y)$ from the condition that (1.8) must be satisfied, i.e. from

$$(AH_x + BH_y)\,\xi_x + (BH_x + CH_y)\,\xi_y = 0. \tag{1.12}$$

The curves $\xi(x, y) = \text{const}$ are the characteristics of equation (1.12). It is always possible to find a subdomain $\delta_1 \subset \delta$,which contains the curve σ and has the property that the function $\varrho(x, y)$, satisfying the equations

$$\xi_x = \varrho(BH_x + CH_y),\; \xi_y = -\varrho(AH_x + BH_y), \tag{1.13}$$

is different from zero everywhere in the domain δ_1.

We conclude, in view of (1.10) and (1.13) that the Jacobian

$$\frac{\partial(\xi, \eta)}{\partial(x, y)} = \varrho(AH_x^2 + 2BH_xH_y + CH_y^2) \neq 0,$$

i.e. that the transformation (1.7) selected in this way is non-singular.

If the equation (1.5) obtained by such a transformation is divided by $AH_x^2 + 2\,BH_xH_y + CH_y^2$, then its dominant part assumes the form

$$\eta^n k(\xi, \eta)\frac{\partial^2 u}{\partial \xi^2} + \frac{\partial^2 u}{\partial \eta^2}, \tag{1.14}$$

where

$$k(\xi, \eta) = \varrho^2\,[x(\xi, \eta)\,, y(\xi, \eta)]\,G[x(\xi, \eta), y(\xi\;\eta)].$$

The image of domain δ_1 in the ξ, η plane is denoted by $\tilde{\delta}$. It should be remembered that n is a positive integer. If $n = 2m + 1$, we can assume without loss of generality that $k > 0$. Indeed, if $k(\xi, \eta) < 0$, then performing the non-singular transformation $\xi_1 = \xi$, $\eta_1 = -\eta$ on expression (1.14) we get

$$\eta_1^{2m+1} k_1(\xi_1, \eta_1)\frac{\partial^2 u}{\partial \xi_1^2} + \frac{\partial^2 u}{\partial \eta_1^2},$$

where

$$k_1(\xi_1, \eta_1) = -k(\xi_1, -\eta_1) > 0.$$

In the domain $\tilde{\delta}$ the characteristic quadratic form

$$\eta^{2m+1} k\mathrm{d}\eta^2 + \mathrm{d}\xi^2,$$

corresponding to the differential expression (1.14) has for its discriminant the function $\eta^{2m+1} k(\xi, \eta)$ which is positive for for $\eta > 0$, zero for $\eta = 0$ and negative for $\eta < 0$. Consequently equation (1.5) is elliptic for $\eta > 0$, parabolic for $\eta = 0$ and hyperbolic for $\eta < 0$, so that when the point $P(x, y)$ crosses the curve σ the equation (1.5) changes its type.

In the case when $n = 2m$, the sign of the discriminant $\eta^{2m} k(\xi, \eta)$ is the same as the sign of the function $k(\xi, \eta)$. Therefore in the considered domain outside the curve σ equation (1.5) is either everywhere elliptic for $k > 0$ or everywhere hyperbolic for $k < 0$.

Consider now what happens when the identity (1.11) holds along the curve σ.

Since A, B and C do not vanish simultaneously it is always possible to find functions $m_1(x, y)$ and $n_1(x, y)$, for which the inequality

$$Am_1^2 + 2Bm_1 n_1 + Cn_1^2 \neq 0. \tag{1.15}$$

holds.

We now choose the functions $\xi = \xi(x, y)$, $\eta = \eta(x, y)$ in such a way as to make the curves $\xi(x, y) = const$, $\eta(x, y) = const$ characteristic curves of the equations

$$(Am_1 + Bn_1)\xi_x + (Bm_1 + Cn_1)\xi_y = 0,$$

$$n_1 \eta_x - m_1 \eta_y = 0,$$

respectively.

Clearly, it is possible to find a section σ_1 of the curve σ and a domain δ, containing the curve σ_1, in which inequality (1.15) holds and at the same time non-vanishing functions $\varrho_1(x, y)$ and $\varrho_2(x, y)$ satisfying the relations

$$\left.\begin{aligned} \eta_x &= \varrho_1 m_1, & \eta_y &= \varrho_1 n_1, \\ \xi_x &= \varrho_2(Bm_1 + Cn_1), & \xi_y &= -\varrho_2(Am_1 + Bn_1). \end{aligned}\right\} \tag{1.16}$$

We conclude from (1.16) that

$$\frac{\partial(\xi, \eta)}{\partial(x, y)} = \varrho_1 \varrho_2 (Am_1^2 + 2Bm_1 n_1 + Cn_1^2) \neq 0.$$

We introduce the notation $H(\xi, \eta) = H[x(\xi, \eta), y(\xi, \eta)]$. Since on σ_1 the directions of the tangents to the curves $H(x, y) =$

$= const$ and $\xi\,(x,\ y) = const$ coincide on account of (1.11) we conclude that $H\,(const,\ \eta) = 0$ and in particular $H\,(0,\ \eta) = 0$. On the other hand, using (1.11) and (1.15) we conclude that

$$H_\xi = \frac{H_x\,\eta_y - H_y\,\eta_x}{\partial\,(\xi,\eta)/\partial\,(x,y)} = \frac{\varrho_1\,(n_1\,H_x - m_1\,H_y)}{\partial\,(\xi,\eta)/\partial\,(x,y)} \neq 0.$$

Hence it follows that in a certain neighbourhood $\delta_1 \subset \delta$ of curve σ_1 we can make the representation

$$H\,(\xi,\eta) = \xi N\,(\xi,\eta). \quad N\,(\xi,\eta) \neq 0. \tag{1.17}$$

After dividing the equation obtained from (1.5) by means of the selected non-singular transformation (1.7) by $A\,\eta_x^2 + 2\,B\eta_x\,\eta_y + C\eta_y^2 = \varrho_1^2\,(Am_1^2 + 2\,Bm_1n_1 + Cn_1^2)$, the dominant part becomes

$$\xi^n\,k\,(\xi,\eta)\,\frac{\partial^2 u}{\partial \xi^2} + \frac{\partial^2 u}{\partial \eta^2}, \tag{1.18}$$

where, in view of (1.17) and the above notation

$$k\,(\xi,\eta) = N^n\,(\xi,\eta)\,\frac{\varrho_2^2}{\varrho_1^2}\,G \neq 0.$$

Let $\bar{\delta}$ be the domain in the ξ, η-plane into which the domain δ_1 in the x, y-plane is transformed.

For $n = 2\,m + 1$, it can be assumed without loss of generality as before, that $k\,(\xi,\ \eta) > 0$. The discriminant of the characteristic quadratic form associated with the differential expression (1.18) has the form $\xi^n k\,(\xi,\ \eta)$ and consequently in the case when $n = 2\,m + 1$ the equation (1.5) in domain $\bar{\delta}$ is elliptic for $\xi > 0$, parabolic for $\xi = 0$ and hyperbolic for $\xi < 0$ in the case when $n = 2\,m$ in domain δ apart from the curve of parabolic degeneracy $\xi = 0$ equation (1.5) is either everywhere elliptic ($k > 0$) or everywhere hyperbolic ($k < 0$).

We now try to simplify further the differential expressions (1.14) and (1.18).

As a result of changing the variables by the formulae

$$z = z\,(\xi,\eta), \quad \zeta = \zeta\,(\xi,\eta) \tag{1.19}$$

the dominant part of the transformed differential expression (1.14) can be written in the form

$$(\eta^n k z_\xi^2 + z_\eta^2) \frac{\partial^2 u}{\partial z^2} + (\eta^n k z_\xi \zeta_\xi + z_\eta \zeta_\eta) \frac{\partial^2 u}{\partial z \, \partial \zeta} + \\ + (\eta^n k \zeta_\xi^2 + \zeta_\eta^2) \frac{\partial^2 u}{\partial \zeta^2} . \tag{1.20}$$

We now choose the transformation (1.19) to be non-singular and to satisfy the conditions:

$$\eta^n k z_\xi \zeta_\xi + z_\eta \zeta_\eta = 0, \tag{1.21}$$

$$\eta^n k z_\xi^2 + z_\eta^2 = \zeta^n (\eta^n k_\xi^2 + \zeta_\eta^2) \operatorname{sgn} k, \tag{1.22}$$

$$\eta^n k \zeta_\xi^2 + \zeta_\eta^2 \neq 0. \tag{1.23}$$

If a transformation of this kind exists, then the differential expression (1.20), after division by the factor $\eta^n k \zeta_\xi^2 + \zeta_\eta^2$, becomes

$$\zeta^{2m+1} \frac{\partial^2 u}{\partial z^2} + \frac{\partial^2 u}{\partial \zeta^2} \text{ for } n = 2m + 1,$$

$$\pm \zeta^{2m} \frac{\partial^2 u}{\partial z^2} + \frac{\partial^2 u}{\partial \zeta^2} \text{ for } n = 2m,$$

where the sign of $\zeta^{2m} \, \partial^2 u / \partial z^2$ is taken to be the same as that of k.

It will now be proved that the non-singular transformation (1.19) satisfying conditions (1.21), (1.22) and (1.23) really exists.

Because of (1.21), (1.22) and (1.23) for $\eta = 0$ we may choose $\zeta = 0$, $z_\eta = 0$, $\zeta_\eta \neq 0$ and from the condition that $\delta (z, \zeta)/\partial/ (\xi, \eta) \neq 0$ near $\eta = 0$, it necessarily follows that $z_\xi \neq 0$. Thus for small $| \eta |$ we must have that $z_\xi \neq 0$, $\zeta_\eta \neq 0$.

We now introduce the function $\varrho (\xi, \eta)$ by the formula

$$z_\xi = \varrho \zeta_\eta \tag{1.24}$$

In order to satisfy (1.21) and (1.22) we must have

$$z_\eta = - \eta^n k \varrho \zeta_\xi, \tag{1.25}$$

$$\zeta = \sigma^{2/n} \eta h, \quad h = |k|^{1/n}. \tag{1.26}$$

From (1.24) and (1.25) we have that

$$(\varrho\zeta_\eta)_\eta + (\eta^n k\varrho\zeta_\xi)_\xi = 0. \tag{1.27}$$

Substituting the expression (1.26) for ζ into (1.27) we get the differential equation which the function $\varrho(\xi, \eta)$ must satisfy:

$$\left[\frac{2}{n}\varrho^{2/n}\eta h\varrho_\eta + \varrho^{(2+n)/2}(h + \eta h_\eta)\right]_\eta +$$
$$+\eta^{n+1}\left[\frac{2}{n}k\varrho^{2/n}h\varrho_\xi + k\varrho^{(2+n)/2}h_\xi\right]_\xi = 0$$

After transforming the dependent variable from ϱ to τ where $\varrho = \tau^{n/(n+2)}$ we see that this equation becomes

$$\left[\frac{2}{n+2}\eta h\tau_\eta + \tau(h + \eta h_\eta)\right]_\eta + \eta^{n+1}\left[\frac{2}{n+2}kh\tau_\xi + k\tau h_\xi\right]_\xi = 0 \tag{1.28}$$

Equation (1.28) is a linear equation of the second order in τ with analytic coefficients, the order of which degenerates along the straight line $\eta = 0$.

In the same way as in the theory of ordinary differential equations we can prove that there is a non-vanishing analytic solution $\tau(\xi, \eta)$ of equation (1.28) at least in some neighbourhood of each point on the line $\eta = 0$. Therefore there exists a function $\varrho(\xi, \eta)$, by means of which, we can, according to formulae (1.24) and (1.25), determine, in some neighbourhood of the selected point of the line $\eta = 0$ the required non-singular transformation (1.19) possessing properties (1.21), (1.22) and (1.23).

In this way in the presence of a curve of parabolic degeneracy σ under condition (1.10) it is always possible to find a non-singular real transformation of the independent variables which in some neighbourhood of a selected point on the curve σ will take equation (1.5) into one of the following simple forms (the old notation for the independent variables is kept):

$$y^{2m+1}\frac{\partial^2 u}{\partial x^2} + \frac{\partial^2 u}{\partial y^2} = F_1\left(x, y, u, \frac{\partial u}{\partial x}, \frac{\partial u}{\partial y}\right),$$

$$y^{2m}\frac{\partial^2 u}{\partial x^2} + \frac{\partial^2 u}{\partial y_2} = F_2\left(x, y, u, \frac{\partial u}{\partial x}, \frac{\partial u}{\partial y}\right),$$

$$y^{2m}\frac{\partial^2 u}{\partial x^2} - \frac{\partial^2 u}{\partial y^2} = F_3\left(x, y, u, \frac{\partial u}{\partial x}, \frac{\partial u}{\partial y}\right).$$

With a view to reducing the differential expression (1.18) to a simpler form we restrict the transformation (1.19) by the conditions:

$$\xi^n k z_\xi \zeta_\xi + z_\eta \zeta_\eta = 0, \tag{1.29}$$

$$\xi^n k z_\xi^2 + z_\eta^2 = z^n (\xi^n k \zeta_\xi^2 + \zeta_\beta^2) \operatorname{sgn} k, \tag{1.30}$$

$$\xi^n k \zeta_\xi^2 + \zeta_\eta^2 \neq 0. \tag{1.31}$$

After such a change of the independent variables the main part of the transformed expression (1.18) assumes, after division by the factor $\xi^n k \zeta_\zeta^2 + \zeta_\eta^2$, one of the following forms:

$$z^{2m+1} \frac{\partial^2 u}{\partial z^2} + \frac{\partial^2 u}{\partial \zeta^2} \text{ for } n = 2m + 1,$$

$$\pm z^{2m} \frac{\partial^2 u}{\partial z^2} + \frac{\partial^2 u}{\partial \zeta^2} \text{ for } n = 2m.$$

By (1.29), (1.30) and (1.31) for $\xi = 0$ we must have $z = 0$, $z_\eta = 0$, $\zeta_\eta \neq 0$. From the requirement that (1.10) is a non-singular transformation $z_\xi \neq 0$.

We shall look for the transformation (1.19) among the analytical solutions of the system:

$$z_\xi = \varrho \zeta_\eta, \tag{1.32}$$

$$z_\eta = -\varrho \xi^n k \zeta_\xi, \tag{1.33}$$

where $\varrho(\xi, \eta)$ is a regular analytic function, different from zero.

From (1.32), (1.33) and (1.30) we get

$$z = \xi \varrho^{2/n} h, \ h = |k|^{1/n}. \tag{1.34}$$

On the basis of (1.33) and (1.34) we conclude that

$$\varrho^{2/n} h = \xi^{n-1} M(\xi, \eta) + N(\xi). \tag{1.35}$$

In what follows it will be assumed that $N(\xi) = \text{const}$, $N \neq 0$ for $n > 1$ and $N = 0$ for $n = 1$. In the last case we also have the natural requirement $M(\xi, \eta) \neq 0$.

Noting (1.34) and (1.35), we get from the condition for the integrability of the system (1.32) and (1.33) the differential equation for determining the function $M(\xi, \eta)$:

$$\frac{\partial^2 M}{\partial \eta^2} + k\xi^n \frac{\partial^2 M}{\partial \xi^2} + \omega\left(\xi, \eta, M, \frac{\partial M}{\partial \xi}, \frac{\partial M}{\partial \eta}\right) = 0. \quad (1.36)$$

In view of the fact that $\xi^{n-1} M + N \neq 0$ for small values of $|\xi|$, the expression $\omega(\xi, \eta, M, \partial M/\partial \xi, \partial M/\partial \eta)$ is an analytic function which is regular at least for small $|\xi|$. Therefore a function $M(\xi, \eta)$ of the required form can be constructed in a neighbourhood of $\xi = 0$, for instance, by solving the Cauchy problem for equation (1.36) with analytic functions for the initial values $M(\xi, \eta_0)$, $\partial M(\xi, \eta)/\partial \eta\,|_{\eta=\eta_0}$.

Substituting the value of ϱ from (1.35) into (1.32) and (1.33) we get a linear system of partial differential equations for determining the required real non-singular transformation of the independent variables (1.19).

Therefore under condition (1.11) there exists a real non-singular transformation of the independent variables which in a neighbourhood of some selected point on the curve of degeneracy transforms equation (1.5) into one of the following canonical forms (the old notation for the independent variables is retained):

$$\frac{\partial^2 u}{\partial x^2} + y^{2m+1} \frac{\partial^2 u}{\partial y^2} = F_4\left(x, y, u, \frac{\partial u}{\partial x}, \frac{\partial u}{\partial y}\right),$$

$$\frac{\partial^2 u}{\partial x^2} + y^{2m} \frac{\partial^2 u}{\partial y^2} = F_5\left(x, y, u, \frac{\partial u}{\partial x}, \frac{\partial u}{\partial y}\right),$$

$$\frac{\partial^2 u}{\partial x^2} - y^{2m} \frac{\partial^2 u}{\partial y^2} = F_6\left(x, y, u, \frac{\partial u}{\partial x}, \frac{\partial u}{\partial y}\right).$$

In this way we have the following canonical forms for linear partial differential equations of second order and of mixed type with two independent variables:

$$y^{2m+1} \frac{\partial^2 u}{\partial x^2} + \frac{\partial^2 u}{\partial y^2} + a \frac{\partial u}{\partial x} + b \frac{\partial u}{\partial y} + cu = f, \quad (1.37)$$

$$\frac{\partial^2 u}{\partial x^2} + y^{2m+1} \frac{\partial^2 u}{\partial y^2} + a \frac{\partial u}{\partial x} + b \frac{\partial u}{\partial y} + cu = f \quad (1.38)$$

and the following canonical forms for linear equations of second order which are parabolically degenerate but of the same type apart from on the curve of degeneracy:

$$y^{2m}\frac{\partial^2 u}{\partial x^2} \pm \frac{\partial^2 u}{\partial y^2} + a\frac{\partial u}{\partial x} + b\frac{\partial u}{\partial y} + cu = f, \tag{1.39}$$

$$\frac{\partial^2 u}{\partial x^2} \pm y^{2m}\frac{\partial^2 u}{\partial y^2} + a\frac{\partial u}{\partial x} + b\frac{\partial u}{\partial y} + cu = f. \tag{1.40}$$

Hence, it is impossible in a region containing a part of the line of degeneracy $y = 0$, to reduce by means of a non-singular transformation of the independent variables any of the equations (1.37), (1.38), (1.39) and (1.40) into another of them or into one of the same type but with a different power of y.

§ 3. Systems of Two First Order Equations

In the present paragraph systems of the following form will be studied:

$$\left.\begin{aligned} a_{11}\frac{\partial u_1}{\partial x} + a_{12}\frac{\partial u_2}{\partial x} + b_{11}\frac{\partial u_1}{\partial y} + b_{12}\frac{\partial u_2}{\partial y} &= f_1(x, y, u_1, u_2),\\ a_{21}\frac{\partial u_1}{\partial x} + a_{22}\frac{\partial u_2}{\partial x} + b_{21}\frac{\partial u_1}{\partial y} + b_{22}\frac{\partial u_2}{\partial y} &= f_2(x, y, u_1, u_2),\end{aligned}\right\} \tag{1.41}$$

where the coefficients a_{ik}, b_{ik} are real functions of the independent variables x, y, and the right sides f_1 and f_2 depend not only on x and y, but also on the unknown functions u_1 and u_2 (in general not linearly).

In what follows it will be assumed that at every point of the domain D at which the system (1.41) is considered the expressions

$$C = b_{11}b_{22} - b_{12}b_{21},\ B = a_{11}b_{22} + b_{11}a_{22} - a_{12}b_{21} - a_{21}b_{12},$$
$$A = b_{11}a_{22} - a_{12}a_{21}$$

never vanish simultaneously. This excludes the case when the system (1.41), generally speaking, is not compatible.

Consider now the so-called characteristic determinant

$$\begin{vmatrix} a_{11} + \lambda b_{11}, & a_{12} + \lambda b_{12} \\ a_{21} + \lambda b_{21}, & a_{22} + \lambda b_{22} \end{vmatrix} = A + 2B\lambda + C\lambda^2.$$

At a point (x, y) of domain D system (1.41) is said to be respectively elliptic, hyperbolic or parabolic according as the function $AC - B^2$ is greater than, less than or equal to zero. We suppose that $C \neq 0$, we can similarly consider the cases when $A = C = 0$, $B \neq 0$ or $C = B = 0$, but $A \neq 0$. System (1.41) may be solved for $\partial u_1/\partial y$ and $\partial u_2/\partial y$ and written in the form

$$\left.\begin{aligned} \frac{\partial u_1}{\partial y} + g_{11}\frac{\partial u_1}{\partial x} + g_{12}\frac{\partial u_2}{\partial x} &= F_1(x, y, u_1, u_2), \\ \frac{\partial u_2}{\partial y} + g_{21}\frac{\partial u_1}{\partial x} + g_{22}\frac{\partial u_2}{\partial x} &= F_2(x, y, u_1, u_2). \end{aligned}\right\} \quad (1.42)$$

We shall not consider here the case when the dominant differential form of system (1.42) is split, i. e. when $g_{12} = g_{21} = 0$.

Assuming that $g_{12} \neq 0$ in domain D, the linear transformation of the unknown functions:

$$v_1 = u_1, \; g_{11} u_1 + g_{12} u_2 = - v_2$$

transforms the system (1.42) to the form:

$$\left.\begin{aligned} \frac{\partial v_1}{\partial y} - \frac{\partial v_2}{\partial x} &= F_1^*(x, y, v_1, v_2), \\ \frac{\partial v_2}{\partial y} + a\frac{\partial v_2}{\partial x} + b\frac{\partial v_2}{\partial x} &= F_2^*(x, y, v_1 v_2). \end{aligned}\right\} \quad (1.43)$$

It must be pointed out that non-singular transformations of the dependent or of the independent variables do not change the type of system (1.41)

The characteristic determinant of the system (1.43) has the form

$$\lambda^2 + b\lambda + a. \quad (1.44)$$

Therefore at points where (1.43) is elliptic the discriminant $\Delta(x, y) = 4a - b^2$ of the quadratic expression (1.44) is greater than zero, at points where the system is hyperbolic it is less than zero and at points where the system is parabolic it is equal to zero.

It is known that by means of a suitably chosen non-singular transformation of the dependent and independent variables, the system (1.43) reduces to the simpler form

$$\frac{\partial u}{\partial \eta} + \frac{\partial v}{\partial \xi} = F_2^{**}(\xi, \eta, u, v),$$

$$\frac{\partial v}{\partial \eta} + a\frac{\partial u}{\partial \xi} = F_1^{**}(\xi, \eta, u, v),$$

where a assumes the values $+1$, -1, or 0 according as the system (1.43) is elliptic, hyperbolic or parabolic in the domain under consideration.

In what follows it will be assumed that the coefficients a and b are analytic functions of variables x and y in the domain D, and that the set of points at which system (1.43) has a parabolic degeneracy is a simple (analytic) curve σ.

In the same way as in the case of one equation of the second order we can show that in some neighbourhood of every point on the curve of parabolic degeneracy σ, it is possible to find canonical forms for the system (1.43).

Suppose that all the partial derivatives of function $\Delta(x, y)$ up to and including order $n-1$ vanish on the curve σ, but that at the same time at least one of the nth derivatives differs from zero. In this case the function $\Delta(x, y)$ can be put in the form

$$\Delta = H^n(x, y)\, G(x, y),$$

where $G(x, y) \neq 0$.

Denote by δ, the subdomain of domain D, containing σ and in which the partial derivatives H_x and H_y do not vanish simultaneously.

There are two possibilities: either

$$H_y^2 + bH_x H_y + aH_x^2 \neq 0, \tag{1.45}$$

$$H_y^2 + bH_x H_y + aH_x^2 = 0. \tag{1.46}$$

Transforming the independent variables according to (1.7) with the non-vanishing jacobian $J = \eta_x \xi_y - \eta_x \xi_y$, and the dependent variables by the linear transformations

$$w_1 = v_1,\ w_2 = \frac{1}{d}\left[(\eta_y + b\eta_x)\,\xi_y + a\eta_x\,\xi_x\right] v_1 + \frac{J}{d}\,v_2$$

under the assumptions:

$$d = \eta_y^2 + b\eta_x\,\eta_y + a\eta_x^2 \neq 0, \tag{1.47}$$

$$(2a\eta_x + b\eta_y)\,\xi_x + (2\eta_y + b\eta_x)\,\xi_\gamma = 0 \tag{1.48}$$

we find that system (1.43) can be rewritten in the form

$$\left.\begin{aligned} &\frac{\partial w_1}{\partial \eta} + \frac{\partial w_2}{\partial \varepsilon} = \varphi_1(\xi, \eta, w_1, w_2),\\ &\frac{\partial w_2}{\partial \eta} - \left\{\frac{a}{d^2} J^2 + \frac{1}{d^2}\left[(\eta_y + b\eta_x)\,\xi_y + \right.\right.\\ &\quad + a\eta_x\,\xi_x]\left[\eta_y(\xi_y + b\xi_x) + \right.\\ &\quad \left.\left. + a\eta_x\,\xi_x\right]\right\}\frac{\partial w_1}{\partial \xi} = \varphi_2(\xi, \eta, w_1, w_2). \end{aligned}\right\} \tag{1.49}$$

In the case when the inequality (1.45) holds we take for $\eta(x, y)$ the function $H(x, y)$, and impose on $\xi(x, y)$ the conditions

$$\xi_x = \varrho\,(2H_y + bH_x), = -\varrho\,(2aH_x + bH_y),$$

where $\varrho \neq 0$. With these variables the system (1.49) becomes:

$$\left.\begin{aligned} &\frac{\partial w_1}{\partial \eta} + \frac{\partial w_2}{\partial \xi} = \varphi_1(\xi, \eta, w_1, w_2),\\ &\frac{\partial w_2}{\partial \eta} - \eta^n k(\xi, \eta)\,\frac{\partial w_1}{\partial \xi} = \varphi_2(\xi, \eta, w_1, w_2), \end{aligned}\right\} \tag{1.50}$$

where

$$k(\xi, \eta) = \varrho^2(\xi, \eta)\,G\,[x(\xi, \eta), y(\xi, \eta)]\,.$$

We consider now the case when the identity (1.46) holds along σ. We then choose the analytic functions $n_1(x, y)$ and $m_1(x, y)$ in such a way as to satisfy:

$$n_1^2 + bm_1\,n_1 + am_1^2 \neq 0. \tag{1.51}$$

The new independent variables ξ and η, are in turn chosen in such a way that $\xi(x, y) = const$ and $\eta(x, y) = const$ become the characteristic equations:

$$(2am_1 + bn_1)\,\xi_x + (2n_1 + bm_1)\,\xi_y = 0,$$
$$n_1\,\eta_x - m_1\,\eta_y = 0,$$

respectively.

It is always possible to find non-vanishing functions $\varrho(\xi, \eta)$ and $\varrho_1(\xi, \eta)$, such that

$$\eta_x = \varrho m_1, \quad \eta_y = \varrho n_1, \quad \xi_x = \varrho_1(2n_1 + bm_1),$$
$$\xi_y = -\varrho_1(2am_1 + bn_1).$$

Since, by (1.51) $d = \varrho^2(n_1^2 + bm_1n_1 + am_1^2) \neq 0$, system (1.43) may reduced to the form of (1.49)

$$\left.\begin{aligned} \frac{\partial w_1}{\partial \eta} + \frac{\partial w_2}{\partial \xi} &= \varphi_1(\xi, \eta, w_1, w_2), \\ \frac{\partial w_2}{\partial \eta} - \xi^n k(\xi, \eta)\frac{\partial w_1}{\partial \xi} &= \varphi_2(\xi, \eta, w_1, w_2), \end{aligned}\right\} \tag{1.52}$$

where

$$k(\xi, \eta) = \frac{\varrho_1^2}{\varrho^2} GN^n(\xi, \eta),$$

$$H[x(\xi, \eta), y(\xi, \eta)] = \xi N(\xi, \eta),$$

and $N(\xi, \eta) \neq 0$.

If we choose the new independent variables $z(\xi, \eta)$ and $\zeta(\xi, \eta)$ to satisfy conditions (1.21), (1.22) and (1.23), and transform the dependent variables by the formulae

$$u = w_1, \quad v = -\frac{J}{d} w_2,$$

where

$$d = \eta^n k\zeta_\xi^2 + \zeta_\eta^2,$$
$$J = z_\eta \zeta_\xi - \zeta_\eta z_\xi,$$

we find that in the neighbourhood of the selected point on curve σ the system (1.50) reduces to one of the following canonical forms

$$\zeta^{2m+1}\frac{\partial u}{\partial z} - \frac{\partial v}{\partial \zeta} = \psi_1(z, \zeta, u, v),$$
$$\frac{\partial u}{\partial \zeta} + \frac{\partial v}{\partial z} = \psi_2(z, \zeta, u, v),$$

or

$$\zeta^{2m}\frac{\partial u}{\partial z} \pm \frac{\partial v}{\partial \zeta} = \psi_1(z, \zeta, u, v),$$

$$\frac{\partial u}{\partial \zeta} + \frac{\partial v}{\partial z} = \psi_2(z, \zeta, u, v)\,.$$

A further simplification of system (1.52) can be brought about in a similar fashion. For this it is sufficient to restrict $z = z(\xi, \eta)$ and $\zeta = \zeta(\xi, \eta)$ by conditions (1.29), (1.30) and (1.31). As a result of a transformation of this kind the system (1.52) assumes the form

$$z^{2m+1}\frac{\partial u}{\partial z} - \frac{\partial v}{\partial \zeta} = \psi_1(z, \zeta, u, v),$$

$$\frac{\partial u}{\partial \zeta} + \frac{\partial v}{\partial z} = \psi_2(z, \zeta, u, v)$$

where ?

$$z^{2m}\frac{\partial u}{\partial z} \pm \frac{\partial v}{\partial \zeta} = \psi_1(z, \zeta, u, v),$$

$$\frac{\partial u}{\partial \zeta} + \frac{\partial v}{\partial z} = \psi_2(z, \zeta, u, v,).$$

§ 4. Linear Systems of Partial Differential Equations of the Second Order with Two Independent Variables

Consider the linear system of partial differential equations of second order

$$A\frac{\partial^2 u}{\partial x^2} + 2B\frac{\partial^2 u}{\partial x\,\partial y} + C\frac{\partial^2 u}{\partial y^2} + A_1\frac{\partial u}{\partial x} + B_1\frac{\partial u}{\partial y} + C_1 u = f, \tag{1.53}$$

where A, B, C, A_1, B_1, C_1 are real square matrices of order n whose elements are functions of the real variables x, y defined in a domain D, $f = (f_1, \ldots, f_n)$ is a given vector and $u = (u_1, \ldots, u_n)$ is the solution vector.

The expression

$$\det |A + 2B\lambda + C\lambda^2|,$$

where λ is a scalar parameter is called the characteristic determinant of the system. This expression is a polynomial of degree $2n$ in λ (Characteristic polynomial).

The direction determined by the equation, $dx/dy = \lambda(x, y)$, where λ is a root of the characteristic polynomial is called the characteristic direction, and the curves determined by the equation $dx/dy = \lambda(x, y)$ the characteristic curves of system (1.53).

If the characteristic polynomial has no real roots at the point (x, y) under consideration then the system (1.53) is said to be elliptic at the point, if all roots of the characteristic polynomial are real and different then the system is said to be hyperbolic in the narrow sense.

In the case in which there are both real and complex among the roots of the characteristic polynomial the system (1.53) is said to be of composite type at the considered point.

If in some set of points in domain D all roots of the characteristic polynomial are real and in the remaining part of domain D there are no real roots then we say that the system (1.53) is of mixed type.

If the coefficients of the second derivatives of u in (1.53) are continuous then at every point where the type of this system changes the multiplicity of the roots of the characteristic polynomial also changes. It is known from elementary algebra that the set of points contained in domain D which satisfy the condition

$$\Delta(x, y) = 0, \tag{1.54}$$

where Δ, the discriminant of the characteristic polynomial, is characterized by the property that on this set at least one of the roots of this polynomial is of at least multiplicity two.

It may happen that equation (1.53) is simultaneously of composite and mixed type in the domain in which it is given. Thus for example the system

$$\left.\begin{aligned} y\frac{\partial^2 u_1}{\partial x^2} + \frac{\partial^2 u_2}{\partial y^2} &= 0 \\ \frac{\partial^2 u_1}{\partial x^2} - \frac{\partial^2 u_1}{\partial y^2} - \frac{\partial^2 u_2}{\partial x^2} + \frac{\partial^2 u_2}{\partial y^2} &= 0, \end{aligned}\right\} \tag{1.55}$$

has the characteristic polynomial

$$(\lambda^2 - 1)(\lambda^2 + y); \tag{1.56}$$

with discriminant $\Delta = y\,(1 + y)$. In the half-plane $y < 0$ the characteristic polynomial has both real $(+1, -1)$ and imaginary $i\sqrt{y}$, $-i\sqrt{y}$ roots whereas in the half-plane $y < 0$ each one of its four roots $+1, -1, \sqrt{(-y)}, -\sqrt{(-y)}$ is real. Consequently the system (1.55) is both of composite and of mixed type in the x, y plane. In this case the locus of the points satisfying condition (1.54) are the straight lines $y = 0$ and $y = -1$, $\lambda = 0$ being the double root on $y = 0$ and $\lambda = 1$, and $\lambda = -1$ being the double roots on $y = -1$.

Systems of the first order and also systems of higher order equations can be classified in similar fashion but we shall not do this here.

CHAPTER 2

THE STUDY OF THE SOLUTIONS OF SECOND ORDER HYPERBOLIC EQUATIONS WITH INITIAL CONDITIONS GIVEN ALONG THE LINES OF PARABOLICITY

§ 1. The Riemann Function for Second Order Hyperbolic Linear Equations

As we saw earlier, a second order linear partial differential equation with two independent variables can, under sufficiently general conditions concerning its coefficient, be reduced in the domain in which it is hyperbolic to the canonical form

$$\frac{\partial^2 u}{\partial x^2} - \frac{\partial^2 u}{\partial y^2} + A(x, y)\frac{\partial u}{\partial x} + \\ + B(x, y)\frac{\partial u}{\partial y} + C(x, y)\,u = F_1(x, y). \tag{2.1}$$

In the characteristic variables $\xi = x + y$, $\eta = x - y$ this equation can be written in the form

$$L(u) = \frac{\partial^2 u}{\partial \xi\, \partial \eta} + a\frac{\partial u}{\partial \xi} + b\frac{\partial u}{\partial \eta} + cu = F, \tag{2.2}$$

where

$$4a = A + B,\ 4b = A - B,\ 4c = C\ \ 4F = F_1.$$

It will be assumed that the coefficients $a(\xi, \eta)$, $b(\xi, \eta)$ are continuously differentiable and that $c(\xi, \eta)$ and $F(\xi, \eta)$ are continuous in the simply connected domain D in which equation (2.2) is given.

It is well known that a very important part is played in the theory of equation (2.2) by the Riemann function which is defined by the following requirements

(1) $v(\xi, \eta)$ is a solution of the adjoint equation

$$M(v) = \frac{\partial^2 v}{\partial \xi \, \partial \eta} - \frac{\partial}{\partial \xi}(av) - \frac{\partial}{\partial \eta}(bv) + cv = 0, \qquad (2.3)$$

(2) the function $v(\xi, \eta)$ assumes on the characteristics $\xi = \xi_1 \; \eta_1 = \eta_1$ the values

$$v(\xi_1, \eta) = \exp\left[\int_{\eta_1}^{\eta} a(\xi_1, \eta_2)\, d\eta_2\right], \quad v(\xi, \eta_1) = \exp\left[\int_{\xi_1}^{\xi} b(\xi_2, \eta_1)\, d\xi_2\right], \qquad (2.4)$$

where (ξ_1, η_1) is an arbitrarily fixed point of the domain D.

Using (2.3) and (2.4) we obtain for the determination of $v(\xi, \eta)$ the integral equation

$$\begin{aligned} v(\xi, \eta) - \int_{\xi_1}^{\xi} b(\xi_2, \eta)\, v(\xi_2, \eta)\, d\xi_2 - \int_{\eta_1}^{\eta} a(\xi\; \eta_2)\, v(\xi, \eta_2)\, d\eta_2 + \\ + \int_{\xi_1}^{\xi} d\xi_2 \int_{\eta_1}^{\eta} c(\xi_2, \eta_2)\, v(\xi_2, \eta_2)\, d\eta_2 = 1. \end{aligned} \qquad (2.5)$$

We now introduce the new function $v_0(\xi, \eta)$, related to function $v(\xi, \eta)$ by the expression

$$\begin{aligned} v(\xi, \eta) = v_0(\xi, \eta) + \int_{\xi_1}^{\xi} v_0(t, \eta)\, b(t, \eta) \exp\left[\int_{t}^{\xi} b(t_1, \eta)\, dt_1\right] dt + \\ + \int_{\eta_1}^{\eta} v_0(\xi, \tau)\, a(\xi, \tau) \exp\left[\int_{\tau}^{\eta} a(\xi, \tau_1)\, d\tau_1\right] d\tau. \end{aligned} \qquad (2.6)$$

By (2.5) and (2.6) we see that $v_0(\xi, \eta)$ is the solution of Volterra's integral equation of the second kind

$$v_0(\xi, \eta) + \int_{\xi_1}^{\xi} dt \int_{\eta_1}^{\eta} k_0(\xi, \eta; t, \tau)\, v_0(t, \tau)\, d\tau = 1, \qquad (2.7)$$

where the kernel $k_0(\xi, \eta; t, \tau)$ is defined by the equation

$$k_0(\xi,\eta;\ t,\tau) = c(t,\ \tau) - b(t,\eta)\, a(t,\tau) \exp\left[\int_\tau^\eta a(t,\tau)\, d\tau_1\right]$$

$$- a(\xi,\tau)\, b(t,\tau)\ \exp\left[\int_t^\xi b(t_1\, \tau)\, dt_1\right]$$

$$+ b(t,\tau) \int_t^\xi c(t_1,\tau)\ \exp\left[\int_t^{t_1} b(t_2,\tau)\, dt_2\right] dt_1 +$$

$$+ a(t,\tau) \int_\tau^\eta c(t,\tau_1)\ \exp\left[\int_\tau^{\tau_1} a(t,\tau_2)\, dc_2\right] d\tau.$$

It is well-known that equation (2.7) has always a unique solution which, for example, can be constructed by means of Picard's method of successive approximations.

In this way under our assumptions concerning the coefficients of equation (2.2) the Riemann function always exists and in addition to variables ξ and η, it depends also on ξ_1 and η_1. Therefore the following notation is used for the Riemann function

$$v = R(\xi, \eta; \xi_1, \eta_1).$$

We have from (2.4)

$$\left.\begin{aligned} \frac{\partial R(\xi_1, \eta; \xi_1, \eta_1)}{\partial \eta} - a(\xi_1, \eta)\, R(\xi_1, \eta; \xi_1, \eta_1) &= 0, \\ \frac{\partial R(\xi, \eta_1; \xi_1, \eta_1)}{\partial \xi} - b(\xi, \eta_1)\, R(\xi, \eta_1; \xi_1, \eta_1) &= 0, \\ R(\xi_1, \eta_1; \xi_1, \eta_1) &= 1; \end{aligned}\right\} \qquad (2.8)$$

$$\left.\begin{aligned} \frac{\partial R(\xi, \eta; \xi, \eta_1)}{\partial \eta_1} + a(\xi, \eta_1)\, R(\xi, \eta; \xi, \eta_1) &= 0, \\ \frac{\partial R(\xi, \eta; \xi_1, \eta)}{\partial \xi_1} + b(\xi_1, \eta)\, R(\xi, \eta; \xi_1, \eta) &= 0, \\ R(\xi, \eta; \xi, \eta) &= 1. \end{aligned}\right\} \qquad (2.9)$$

For every function $u(\xi_1, \eta_1)$ which is differentiable up to a suitable order in domain D we have the identity

$$\frac{\partial^2}{\partial \xi_1 \partial \eta_1}[u(\xi_1, \eta_1) R(\xi_1, \eta_1; \xi, \eta)] - R(\xi_1, \eta_1; \xi, \eta) L(u) = \\ = \frac{\partial}{\partial \xi_1}\left[u\left(\frac{\partial R}{\partial \eta_1} - aR\right)\right] + \frac{\partial}{\partial \eta_1}\left[u\left(\frac{\partial R}{\partial \xi_1} - bR\right)\right]. \tag{2.10}$$

If the derivatives contained in identity (2.10) are required to be continuous, then from (2.10) and (2.8) we get by means of integrating with respect to variables ξ_1 and η_1 between the limits $\xi_0 \leqslant \xi_1 \leqslant \xi$, $\eta_0 \leqslant \eta_1 \leqslant \eta$,

$$u(\xi, \eta) = u(\xi_0, \eta_0) R(\xi_0, \eta_0; \xi, \eta) + \\ + \int_{\xi_0}^{\xi} R(\xi_1, \eta_0; \xi, \eta)\left[\frac{\partial u(\xi_1, \eta_0)}{\partial \xi_1} + b(\xi_1, \eta_0) u(\xi_1, \eta_0)\right] d\xi_1 + \\ + \int_{\eta_0}^{\eta} R(\xi_0, \eta_1; \xi, \eta)\left[\frac{\partial u(\xi_0, \eta_1)}{\partial \eta_1} + a(\xi_0, \eta_1) u(\xi_0, \eta_1)\right] d\eta_1 + \\ + \int_{\xi_0}^{\xi} d\xi_1 \int_{\eta_0}^{\eta} R(\xi_1, \eta_1; \xi, \eta) L[u(\xi_1, \eta_1)] d\eta_1, \tag{2.11}$$

where (ξ_0, η_0) is an arbitrary point of domain D.

In particular, putting $u(\xi, \eta) = R(\xi_0, \eta_0; \xi, \eta)$, in (2.11) and using (2.9) we have

$$\int_{\xi_0}^{\xi} d\xi_1 \int_{\eta_0}^{\eta} R(\xi_1, \eta_1; \xi, \eta) L[R(\xi_0, \eta_0; \xi_1, \eta_1)] d\eta_1 = 0,$$

whence it follows immediately that with respect to its last pair or arguments ξ_1, η_1, the Riemann function $R(\xi, \eta; \xi_1, \eta_1)$ is a solution of the homogeneous equations

$$L(u) = 0. \tag{2.12}$$

In this chapter by the solution of a hyperbolic equation in the considered domain we mean the function which satisfies the equation and has continuous derivatives of all orders occurring in the equation.

It is evident from the above properties of the Riemann function that one of the special solutions for the homogeneous equation (2.2) is of the form

$$u_0(\xi,\eta) = \int_{\xi_0}^{\xi} d\xi_1 \int_{\eta_0}^{\eta} R(\xi_1,\eta_1;\ \xi,\eta)\, F(\xi_1,\eta_1)\, d\eta_1.$$

Assuming now that $u(\xi,\eta)$ is the solution of equation (2.2) we get from identity (2.11) after integrating by parts

$$\begin{aligned} u(\xi,\eta) = {} & R(\xi,\eta_0;\ \xi,\eta)\,u(\xi,\eta_0) + R(\xi_0,\eta;\ \xi,\eta)\,u(\xi_0,\eta) - \\ & - R(\xi_0,\eta_0;\ \xi,\eta)\,u(\xi_0\eta_0) + \\ & + \int_{\xi_0}^{\xi}\left[b(t,\eta_0)\,R(t,\eta_0;\xi,\eta) - \frac{\partial R(t,\eta_0;\xi,\eta)}{\partial t}\right]u(t,\eta_0)\,dt + \\ & + \int_{\eta_0}^{\eta}\left[a(\xi_0,\tau)\,R(\xi_0,\tau;\ \xi,\eta) - \frac{\partial R(\varepsilon_0,\tau;\xi,\eta)}{\partial \tau}\right]u(\xi_0,\tau)\,d\tau + \\ & + \int_{\xi_0}^{\xi} dt \int_{\eta_0}^{\eta} R(t,\tau;\ \xi,\eta)\,F(t,\tau)\,d\tau. \end{aligned} \tag{2.13}$$

Noting the above properties of the Riemann function we can conclude immediately that the right side of formula (2.13) when $u(\xi_0,\ \eta)$ and $u(\eta_0,\ \xi)$ are replaced in it by arbitrary functions of η and ξ respectively and $u(\xi_0,\ \eta_0)$ by an arbitrary constant, furnishes the solution to equation (2.2).

In particular, (2.13) automatically implies the existence, uniqueness and stability of the solution to the characteristic problem of Goursat: find the solution $u(\xi_0,\ \eta_0)$ of equation (2.2) satisfying the conditions $(u\ (\xi,\ n))$

$$u(\xi,\eta_0) = \varphi(\xi),\ u(\xi_0,\eta) = \psi(\eta),\ \varphi(\xi_0) = \psi(\eta_0),$$

where φ and ψ are given functions.

The solution is of the form

$$u(\xi,\eta) = R(\varepsilon,\eta_0;\ \xi,\eta)\,\varphi(\xi) + R(\xi_0,\eta;\ \xi,\eta)\,\psi(\eta) - R(\xi_0,\eta_0;\ \xi,\eta)\,\varphi(\xi_0) + \int_{\eta_0}^{\eta}\left[a(\xi_0,\tau)\,R(\xi_0,\tau;\ \xi,\eta) - \frac{\partial R(\xi_0,\tau;\ \xi,\eta)}{\partial\tau}\right]\psi(\tau)\,\mathrm{d}\tau + \int_{\xi_0}^{\xi}\left[b(t,\eta_0)\,R(t,\eta_0;\,\eta) - \frac{\partial R(t,\eta_0;\ \xi,\eta)}{\partial t}\right]\varphi(t)\,\mathrm{d}t + \int_{\xi_0}^{\xi}\mathrm{d}t\int_{\eta_0}^{\eta}R(t,\tau;\ \xi,\eta)\,F(t,\tau)\,\mathrm{d}\tau.$$

By means of the Riemann functions we can express the solution of the Cauchy problem for equation (2.2) in terms of integrals.

Denote by σ, a simple closed Jordan curve with continuously varying normal which is contained in the domain D in which equation (2.2) is given and possesses the additional property that it never touches the characteristics of equation (2.2).

Assume that the characteristics $\xi_1 = \xi$, $\eta_1 = \eta$ of equation (2.2) starting from point $P(\xi, \eta)$ intersect σ in points Q' and Q respectively, and that G is the finite domain enclosed by curve QQ' of curve σ and the characteristics PQ and PQ'.

For arbitrary functions $u(\xi_1, \eta_1)$ an $v(\xi_1, \eta_1)$ which are continuously twice differentiable in the domain D we have the identity:

$$2\,[vL(u) - uM(v)] = \frac{\partial}{\partial\eta_1}\left(\frac{\partial u}{\partial\xi_1}\,v - u\,\frac{\partial v}{\partial\xi_1} + 2buv\right) + \frac{\partial}{\partial\xi_1}\left(\frac{\partial u}{\partial\eta_1}\,v - u\,\frac{\partial v}{\partial\eta_1} + 2auv\right). \tag{2.14}$$

Integrating (2.14) over domain G, we have

$$2\iint_G [vL(u) - uM(v)]\,\mathrm{d}\xi_1\,\mathrm{d}\eta_1 = \int_{PQ+\sigma+Q'P}\left(\frac{\partial u}{\partial\eta_1}\,v - u\,\frac{\partial v}{\partial\eta_1} + 2auv\right)\mathrm{d}\eta_1 - \left(\frac{\partial u}{\partial\xi_1}\,v - u\,\frac{\partial v}{\partial\xi_1} + 2ubv\right)\mathrm{d}\xi_1. \tag{2.15}$$

In the case in which $u(\xi_1, \eta_1)$ is the solution of (2.2) and $v(\xi_1, \eta_1) = R(\xi_1, \eta_1; \xi, \eta)$ integrating by parts and using the properties of the Riemann function we get from (2.15)

$$\begin{aligned} u(P) = \frac{1}{2} u(Q) R(Q,P) + \frac{1}{2} u(Q') R(Q'.P) + \\ + \iint_G F(P') R(P',P) \, d\xi_1 \, d\eta_1 + \frac{1}{2} \int_\sigma \Big\{ \frac{\partial u(P')}{\partial N'} R(P',P) - \\ - u(P') \frac{\partial R(P',P)}{\partial N'} + 2 \Big[a(P') \frac{\partial \xi_1}{\partial N} + \\ + b(P') \frac{\partial \eta_1}{\partial N} \Big] R(P',P) \, u(P') \Big\} \, ds, \end{aligned} \tag{2.16}$$

where N is the normal of curve σ at point $P'(\xi_1, \eta_1)$, directed inside the domain G, and $\partial/\partial N' = (\partial\xi_1/\partial N)\, \partial/\partial\eta_1 + (\partial\eta_1/\partial N)\, \partial/\partial\xi_1$.

From given values of the solution $u(P)$ and its derivative on σ where the direction $\vec{l}$ is nowhere the same as the direction of the tangent of σ it is always possible to calculate the values of $\partial u/\partial N'$ along σ.

Therefore at every point $P(\xi, \eta)$ with the property that the characteristics starting from it intersect (of course, each only once) the curve σ in points Q and Q', the formula (2.16) furnishes the solution (ξ, η) to equation (2.2) by means of the known values of u and $\partial u/\partial l$ given on curve QQ' of curve σ, i. e. it solves the Cauchy problem.

The form of the integral equation which serves for determining the Riemann function shows that when the continuous differentiability of $a(\xi, \eta)$, $b(\xi, \eta)$ and the continuity of $c(\xi, \eta)$ are assumed, the Riemann function has a continuous second mixed derivative. Noting this circumstance, we conclude from (2.16) that from the continuity of the curvature of σ, the twice continuous differentiability of $\tau = u/\sigma$ and the continuous differentiability of $\nu = \partial u/\partial l/\sigma$ it follows that the solution of the Cauchy problem is twice continuously differentiable.

It is clear from formula (2.16) and from the way it was obtained that under the assumptions listed above the Cauchy problem always allows a solution which is also unique and stable.

By demanding that curve σ and the initial conditions should be suitably smooth the investigation of the Cauchy problem (and also that of Goursat) can be reduced to the case when either equation (2.2) is homogencous of the type (2.12) and the initial conditions are inhomogeneous, or the initial conditions are homogeneous and the equation is inhomogeneous.

§ 2. A Class of Hyperbolic Systems of Second Order Linear Equations

The results obtained in the preceding paragraph can be easily generalized to include systems of linear second order partial differential equations of the following form

$$\frac{\partial^2 u_i}{\partial x^2} - \frac{\partial^2 u_i}{\partial y^2} + \sum_{k=1}^{n} \left[A_{ik}(x, y) \frac{\partial u_k}{\partial x} + B_{ik}(x, y) \frac{\partial u_k}{\partial y} + \right.$$
$$\left. + C_{ik}(x, y)\, u_k \right] = F_i(u, y), \quad i = 1, 2, \ldots, n, \tag{2.18}$$

where $A_{ik} = A_{ki}$, $B_{ik} = B_{ki}$, $C_{ik} = C_{ki}$.

According to the classification adopted in the first chapter the system (2.17) is hyperbolic. A general hyperbolic linear system of second order clearly cannot be reduced to the form of (2.17).

When investigating this system it is more convenient to use the matrix notation:

$$L(u) = \frac{\partial^2 u}{\partial x^2} - \frac{\partial^2 u}{\partial y^2} + A \frac{\partial u}{\partial x} + B \frac{\partial u}{\partial y} + Cu = F, \tag{2.12}$$

where $A = ||A_{ik}||$, $B = ||B_{ik}||$, $C = ||C_{ik}||$ are given symmetric real square matrices of order $F = (F_1, \ldots, F_n)$ is a given vector and $u = (u_1, \ldots, u_n)$ is the solution vector which is to be found.

System (2.18) in the characteristic variables $\xi = x + y$, $\eta = x - y$ becomes

$$L(u) = \frac{\partial^2 u}{\partial \xi \, \partial \eta} + a \frac{\partial u}{\partial \xi} + b \frac{\partial u}{\partial \eta} + cu = f, \tag{2.19}$$

where $4a = A + B$, $4b = A - B$, $4c = C$, $4f = F$.

It will be assumed in what follows that matrices a and b have continuous first derivatives and that matrix c and vector f are continuous in the domain D in which the system (2.19) is given.

The adjoint of (2.19) plays an important part in the theory dealing with this system, it being defined by

$$M(v) = \frac{\partial^2 v}{\partial \xi \, \partial \eta} - \frac{\partial}{\partial \xi}(va) - \frac{\partial}{\partial \eta}(vb) + vc = 0. \qquad (2.20)$$

Since, in general, it is not true that $va = av$, $vb = bv$, $vc = cv$, the orders of the factors on the left side of (2.20) must be preserved.

We shall say that the matrix $R(\xi, \eta; \xi_1, \eta_1) = \| R_{ik}(\xi, \eta; \xi_1, \eta_1) \|$ is the Riemann matrix for system (2.19) if the following conditions are satisfied: (1) each of its rows is a solution of system (2.20) in the variables ξ and η (2) for matrices $R(\xi_1, \eta; \xi_1, \eta_1)$ and $R(\xi, \eta_1; \xi_1, \eta_1)$ we have the identities

$$\left.\begin{aligned} &\frac{\partial R(\xi_1, \eta; \xi_1, \eta_1)}{\partial \eta} - R(\xi_1, \eta; \xi_1, \eta_1)\, a(\xi_1, \eta) = 0, \\ &\frac{\partial R(\xi, \eta_1; \xi_1, \eta_1)}{\partial \xi} - R(\xi, \eta_1; \xi_1, \eta_1)\, b(\xi, \eta_1) = 0, \\ &\qquad\qquad R(\xi_1, \eta_1; \xi_1, \eta_1) = E, \end{aligned}\right\} \qquad (2.21)$$

where E is the diagonal unit matrix of order n.

It is easily shown that under these conditions the matrix $R(\xi, \eta; \xi_1, \eta_1)$ is uniquely determined. In fact, conditions (1) and (2) lead to the system of integral equations of the Volterra type.

$$\begin{aligned} R(\xi, \eta; \xi_1 \eta_1) &- \int_{\xi_1}^{\xi} R(\xi_2, \eta; \xi_1, \eta_1)\, b(\xi_2, \eta)\, d\xi_2 - \\ &- \int_{\eta_1}^{\eta} R(\xi, \eta_2; \xi_1, \eta_1)\, a(\xi, \eta_2)\, d\eta_2 + \qquad (2.22) \\ &+ \int_{\xi_1}^{\xi} d\xi_2 \int_{\eta_1}^{\eta} R(\xi_2, \eta_2; \xi_1 \eta_1)\, c(\xi_2, \eta_2)\, d\eta_2 = E. \end{aligned}$$

for the determination of the function $R(\xi, \eta; \xi_1, \eta_1)$.

There is no difficulty involved in proving the existence and the uniqueness of the solution of equation (2.22).

We shall now establish the basic properties of the Riemann matrix. Since as a function of η, $R(\xi_1, \eta; \xi_1, \eta_1)$ is the matrix of fundamental solutions of the system of ordinary differential solutions of the system of ordinary differential equations $\partial R/\partial\eta - Ra = 0$, which satisfies the initial conditions $R(\xi_1, \eta_1; \xi_1, \eta_1) = E$, the following identity holds

$$\det R(\xi_1, \eta; \xi_1, \eta_1) = \exp\left[\sum_{k=1}^{n} \int_{\eta_1}^{\eta} a_{kk}(\xi_1, \eta_2)\, d\eta_2\right]. \quad (2.23)$$

As a result of the first of the identities (2.21) we find that for an arbitrary continuously differentiable matrix $v(\tau_1)$ we have the relation

$$\frac{\partial}{\partial r_1}[R(\xi, r_1; \xi, \eta)\, v(r_1)]$$
$$- R(\xi, \tau_1; \xi, \eta)\left[\frac{dv(\tau_1)}{d\tau_1} + a(\xi, \tau_1)\, v(\tau_1)\right] = 0;$$

whence we get after integrating with respect to τ_1,

$$v(\eta) = R(\xi_1, \eta_1; \xi, \eta)\, v(\eta_1) +$$
$$+ \int_{\eta_1}^{\eta} R(\xi, \tau_1; \xi, \eta)\left[\frac{dv(\tau_1)}{d\tau_1} + a(\xi, \tau_1)\, v(\tau_1)\right] d\tau_1.$$

If we take $v(\tau) = R(\xi, \eta_1; \xi, \tau)$, in the last identity, we have

$$\int_{\eta_1}^{\eta} R(\xi, \tau_1; \xi\ \eta)\left[\frac{\partial R(\xi, \eta_1; \xi, \tau_1)}{\partial \tau_1} + a(\xi, \tau_1)\, R(\xi, \eta_1; \xi, \tau_1)\right] d\tau_1 = 0.$$

Hence, using (2.23) we get

$$\frac{\partial R(\xi, \eta; \xi, \eta_1)}{\partial \eta_1} + a(\xi, \eta_1)\, R(\xi, \eta; \xi. \eta_1) = 0, \quad (2.24)$$

In similar fashion we conclude that

$$\frac{\partial R(\xi, \eta; \xi_1, \eta)}{\partial \xi_1} + b(\xi_1, \eta)\, R(\xi, \eta; \xi_1, \eta) = 0. \quad (2.25)$$

By direct verification it is easy to see the validity of the following identity:

$$\frac{\partial^2}{\partial t\,\partial \tau}[R(t,\tau;\xi,\eta)\,u(t,\tau)] - R(t,\tau;\xi.\eta)\,L(u) =$$

$$= \frac{\partial}{\partial t}\left[\left(\frac{\partial R}{\partial \tau} - Ra\right)u\right] + \frac{\partial}{\partial \tau}\left[\left(\frac{\partial R}{\partial t} - Rb\right)u\right],$$

where u is an arbitrary square matrix of order n. From this we get after integrating and using (2.21), the formula

$$u(\xi,\eta) = Q(\xi_0,\eta_0;\xi,\eta)\,u(\xi_0,\eta_0) +$$

$$+ \int_{\xi_0}^{\xi} \mathrm{d}\xi_1 \int_{\eta_0}^{\eta} R(\xi_1,\eta_1;\xi,\eta)\,L[u(\xi_1,\eta_1)]\,\mathrm{d}\eta_1 +$$

$$+ \int_{\xi_0}^{\xi} R(\xi_1,\eta_0;\xi,\eta)\left[\frac{\partial u(\xi_1,\eta_0)}{\partial \xi_1} + b(\xi_1,\eta_0)\,u(\xi_1,\eta_0)\right]\mathrm{d}\xi_1 +$$

$$+ \int_{\eta_0}^{\eta} R(\xi_0,\eta_1;\xi,\eta)\left[\frac{\partial u(\xi_0,\eta_1)}{\partial \eta_1} + a(\xi_0,\eta_1)\,u(\xi_0,\eta_1)\right]\mathrm{d}\eta_1.$$

If we put here $u(\xi,\ \eta) = R(\xi_0,\ \eta_0;\ \xi,\ \eta)$, and use equations (2.24) and (2.25) we find that

$$\int_{\xi_0}^{\xi} \mathrm{d}\xi_1 \int_{\eta_0}^{\eta} R(\xi_1,\eta_1;\xi,\eta)\,L(\xi_0,\eta_0;\xi_1,\eta_1)]\,\mathrm{d}\eta_1 = 0. \quad (2.27)$$

From (2.23), due to (2.27) it follows immediately that each column of the Riemann matrix $R(\xi,\ \eta;\ \xi_1,\ \eta_1)$ considered as a function of the second pair of arguments ξ_1, η_1 is a solution of the homogeneous system

$$L(u) = 0. \quad (2.28)$$

On account of this circumstance we conclude that the vector

$$u_0(\xi,\eta) = \int_{\xi_0}^{\xi} \mathrm{d}t \int_{\eta_0}^{\eta} R(t,\tau;\eta)\,f(t,\tau)\,\mathrm{d}\tau$$

is a special solution of system (2.19). It is clear from this that the investigation of a number of linear problems concerned with the non-homogeneous system (2.19) can be reduced to the investigation of suitably chosen linear problems for the homogeneous system (2.28).

On the assumption that $u(\xi, \eta)$ is a solution of system (2.28) we get by integrating by parts from (2.26).

$$\begin{aligned} u(\xi,\eta) &= R(\xi,\eta_0;\xi,\eta)\,u(\xi,\eta_0) + R(\xi_0,\eta;\xi,\eta)\,u(\xi_0,\eta) - \\ &\quad - R(\xi_0,\eta_0;\xi,\eta)\,u(\xi_0,\eta_0) - \\ &\quad - \int_{\xi_0}^{\xi}\left[\frac{\partial R(t,\eta_0;\xi,\eta)}{\partial t} - b(t,\eta_0)\,R(t,\eta_0;\xi,\eta)\right]u(t,\eta_0)\,dt - \\ &\quad - \int_{\eta_0}^{\eta}\left[\frac{\partial R(\xi_0,\tau;\xi,\eta)}{\partial \tau} - a(\xi_0,\tau)\,R(\xi_0,\tau;\xi,\eta)\right]u(\xi_0,\tau)\,d\tau. \end{aligned} \tag{2.29}$$

Formula (2.29) furnishes the solution to the following Goursat problem: Find the solution $u(\xi, \eta)$, of system (2.28) which assumes on the characteristics $\xi = \xi_0$, $\eta = \eta_0$ the given values

$$u(\xi,\eta_0) = \varphi(\xi),\ u(\xi_0,\eta) = \psi(\eta),\ \varphi(\xi_0) = \psi(\eta_0),$$

where $\varphi(\xi)$ and $\psi(\eta)$ are fixed vectors.

With a view to finding further applications for the Riemann matrix we note the following vector analogue of identity (2.14)

$$\begin{aligned} R(\xi_1,\eta_1;\xi,\eta)\,L[u(\xi_1\eta_1)] - M[R(\xi_1,\eta_1;\xi,\eta)]\,u(\xi_1,\eta_1) &= \\ = \frac{1}{2}\frac{\partial}{\partial\eta_1}\left[R\frac{\partial u}{\partial\xi_1} - \frac{\partial R}{\partial\xi_1}u + 1Rbu\right] &+ \\ + \frac{1}{2}\frac{\partial}{\partial\xi_1}\left[R\frac{\partial u}{\partial\eta_1} - \frac{\partial R}{\partial\eta_1}u + 2Rau\right]&. \end{aligned} \tag{2.30}$$

where on this occassion u is a vector.

Assuming that u is a solution to system (2.28), and integrating over the domain G, considered in the preceding paragraph, we get from identity (2.30) the formula

$$\begin{aligned} u(P) &= \frac{1}{2}R(Q,P)\,u(Q) + \frac{1}{2}R(Q'.P)\,u(Q') + \\ &+ \frac{1}{2}\int_{\sigma}\left[R\frac{\partial u}{\partial N'} - \frac{\partial R}{\partial N'}u + 2R\left(a\frac{\partial\xi_1}{\partial N} + b\frac{\partial\eta_1}{\partial N}\right)u(P')\right]ds, \end{aligned} \tag{2.31}$$

giving the solution to the Cauchy problem

$$u\Big|_{\sigma} = \tau, \quad \frac{\partial u}{\partial l}\Big|_{\sigma} = \nu,$$

where $\tau = (\tau_1, \ldots, \tau_n)$, $\nu = \nu_1, \ldots, \nu_n)$ are given vectors.

§ 3. The Cauchy Problem for Hyperbolic Equations with Given Initial Conditions on the Line of Parabolic Degeneracy

In the present paragraph we shall be dealing with the equations

$$y^m \frac{\partial^2 u}{\partial x^2} - \frac{\partial^2 u}{\partial y^2} + a\frac{\partial u}{\partial y} + b\frac{\partial u}{\partial y} + cu = 0, \tag{2.32}$$

$$\frac{\partial^2 u}{\partial x^2} - y^m \frac{\partial^2 u}{\partial y^2} + a\frac{\partial u}{\partial x} + b\frac{\partial u}{\partial y} + cu = 0 \tag{2.33}$$

which are of hyperbolic type on the half-plane $y > 0$, with parabolic degeneracy along the straight line $y = 0$.

According to the results obtained in § 2 of the preceding chapter, a general linear hyperbolic equation of second order with certain conditions on its coefficients can be reduced to types (2.32) or (2.33) with positive integer values m.

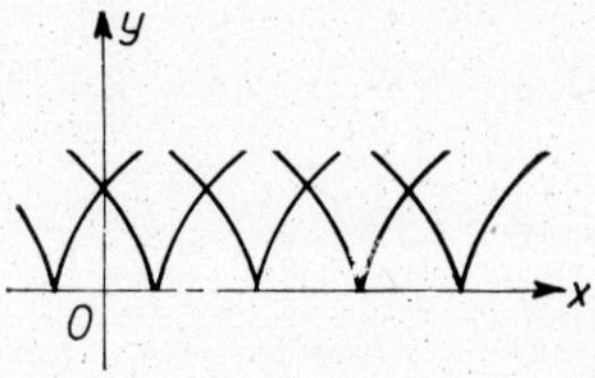

FIG. 1

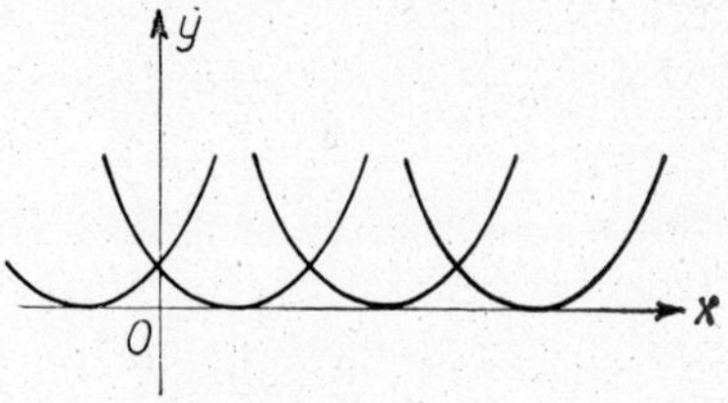

FIG. 2

In what follows we shall assume that the power m of variable y in equations (2.32) and (2.33) is a positive real number.

The family of curves

$$(x - c_0)^2 - \left(\frac{2}{m+2}\right)^2 y^{m+2} = 0, \tag{2.34}$$

$$(x - c_0)^2 - \left(\frac{2}{2-m}\right)^2 y^{2-m} = 0, \; m \neq 2, \; c_0 = \text{const}, \tag{2.35}$$

consists of the characteristics of equations (2.32) and (2.33) respectively, and the line of degeneracy $y = 0$ is the envelope for family (2.34) and the cusp locus for family (2.35) for $m < 2$ (cf. Figs. 1 and 2, respectively).

The curve $y = 0$ together with the curves (2.35) for $m > 2$ and with the curves $(y - e^{c_0+x})\ (y - e^{c_1-x}) = 0$ for $m = 2$ constitute the family of characteristics for equation (2.33).

In the characteristic variables

$$\xi = x + \frac{2}{m+2}\, y^{(m+2/2)}, \quad \eta = x - \frac{2}{m+2}\, y^{(m+2/2)} \tag{2.36}$$

equation (2.32) can be written in the following manner:

$$\frac{\partial^2 u}{\partial \xi\, \partial \eta} + \left[a \left(\frac{4}{m+2} \right)^{m-2/m+2} (\xi - \eta)^{(2-m)/(2+m)} + \right.$$

$$\left. + b \left(\frac{4}{m+2} \right)^{-2/(m+2)} (\xi - \eta)^{2/(m+2)} - \frac{m}{2} \right) \frac{1}{m+2} \, \frac{1}{\xi - \eta} \, \frac{\partial u}{\partial \xi} +$$

$$+ \left[a \left(\frac{4}{m+2} \right)^{(m-2)/(m+2)} (\xi - \eta)^{(2-m)/(2+m)} - \right.$$

$$\left. - b \left(\frac{4}{m+2} \right)^{-2/(m+2)} (\xi - \eta)^{2/(m+2)} + \frac{m}{2} \right] \frac{1}{m+2} \, \frac{1}{\xi - \eta} \, \frac{\partial u}{\partial \eta} +$$

$$+ \frac{c}{2} \left(\frac{4}{m+2} \right)^{2m/(m+2)} (\xi - \eta)^{-2m/(m+2)}\, u = 0. \tag{2.37}$$

Transformation (2.36) is non-singular for $y > 0$, and it maps the half-plane $y > 0$ onto the half-plane $\xi > \eta$. The straight line $y = 0$, i. e. $\xi - \eta = 0$, is the singular line of this transformation.

The assumption that coefficients $a\,(x,\ y)$, $b\,(x,\ y)$ are continuously differentiable and coefficient $c\,(x,\ y)$ is continuous in equation (2.32) for the half-plane $y \geqslant 0$ implies that the coefficients standing beside $\partial u/\partial \xi$, $\partial u/\partial \eta$ and u in equation (2.37) possess the same properties in the half-plane $\xi > \eta$, and become infinite on the straight line $\xi = \eta$.

While the Riemann function and the solution to the Cauchy and Goursat problem for equation (2.32) are constructed for the half-plane $\xi \geqslant \eta + \varepsilon$, where ε is an arbitrary positive number by the standard method as was done in § 1 of this chapter, the Cauchy problem with the initial conditions given on the curve of parabolic degeneracy $y = 0$ requires a special investigation.

Consider first the case when coefficients a, b and c in equation (2.32) vanish identically, i.e.

$$y^m \frac{\partial^2 u}{\partial x^2} - \frac{\partial^2 u}{\partial y^2} = 0. \tag{2.38}$$

Our aim is to study the Cauchy problem in the following sense: find the solution $u(x, y)$ to equation (2.38) which is continuous together with its derivatives up to (and including) the second derivative for $y > 0$ and which satisfies on some section of the axis $y = 0$, for example on the section between $A(0, 0)$ $B(1, 0)$ the conditions

$$\lim_{y \to +0} u(x, y) = \tau(x), \tag{2.39}$$

$$\lim_{y \to +0} \frac{\partial u(x, y)}{\partial y} = \nu(x), \tag{2.40}$$

where $\tau(x)$ and $\nu(x)$ are given functions such that $\tau(x)$ is continuous together with its derivatives up to the second, and $\nu(x)$ and its first derivative are continuous.

In the characteristic variables (2.36) equation (2.38) becomes an Euler—Darboux equation:

$$\frac{\partial^2 u}{\partial \xi\, \partial \eta} - \frac{m}{2(m+2)} \frac{1}{\xi - \eta} \frac{\partial u}{\partial \xi} + \frac{m}{2(m+2)} \frac{1}{\xi - \eta} \frac{\partial u}{\partial \eta} = 0, \tag{2.41}$$

and the conditions (2.39), (2.40) assume the form

$$\left.\begin{aligned} \lim_{\xi - \eta \to +0} u(\xi, \eta) &= \tau(\xi), \\ \lim_{\xi - \eta \to +0} (\xi - \eta)^{m/(m+2)} \left(\frac{m+2}{4}\right)^{m/(m+2)} \left(\frac{\partial u}{\partial \xi} - \frac{\partial u}{\partial \eta}\right) &= \nu(\xi). \end{aligned}\right\} \tag{2.42}$$

? The integral equation (2.5) for the Riemann function $R(\xi_1, \eta_1; \xi, \eta)$, $\xi_1 > \eta_1$, $\xi_1, \geqslant \eta_1 \geqslant \eta$, becomes significantly simpler:

$$\begin{aligned} R(\xi_1, \eta_1; \xi, \eta) &- \frac{m}{2(m+2)} \int_{\xi_1}^{t} \frac{S(\xi_2, \eta_1; \xi, \eta)}{\xi_2 - \eta_1} \, d\xi_2 + \\ &+ \frac{m}{2(m+2)} \int_{\eta}^{\eta} \frac{R(\xi_1, \eta_2; \xi, \eta)}{\xi_1 - \eta_2} \, d\eta_2 = 1. \end{aligned} \tag{2.43}$$

When the method of successive approximations is applied to solve integral equation (2.43) we get the expression:

$$R(\xi_1, \eta_1; \xi, \eta) = [(\xi_1 - \eta)(\xi - \eta_1)]^{-m/(2m+4)} (\xi_1 - \eta_1)^{m/(m+2)} \times \\ \times F\left[\frac{m}{2m+4}, \frac{m}{2m+4}, 1, \frac{(\xi - \xi_1)(\eta_1 - \eta)}{(\xi_1 - \eta)(\xi - \eta_1)}\right], \tag{2.44}$$

where F is the hypergeometric function.

Denote by G the domain enclosed by the segment $Q\ (\eta + \varepsilon, \eta)$ $Q'\ (\xi, \xi - \varepsilon)$ of the straight line $\xi_1 = \eta_1 + \varepsilon$, $\varepsilon > 0$ and the characteristics QP: $\eta_1 = \eta$, $Q'P$: $\xi_1 = \xi$ (Fig. 3).

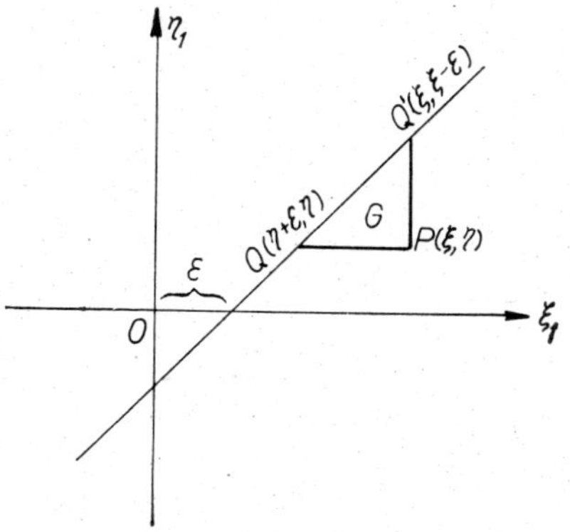

FIG. 3

According to (2.16) the following identity is true for any twice continuously differentiable solution $u(\xi, \eta)$ of equation (2.41):

$$\begin{aligned} u(\xi, \eta) &= \frac{1}{2} u(\eta + \varepsilon, \eta) R(\eta + \varepsilon, \eta; \xi, \eta) + \\ &+ \frac{1}{2} u(\xi, \xi - \varepsilon) R(\xi, \xi - \varepsilon; \xi, \eta) - \\ &- \frac{1}{2} \int_{\eta+\varepsilon}^{\xi} u(\xi_1, \xi_1 - \varepsilon) \left[\frac{\partial R(\xi_1 \eta_1; \xi, \eta)}{\partial \xi_1} - \frac{\partial R(\xi_1, \eta_1; \xi, \eta)}{\partial \eta_1} - \right. \\ &\left. - \frac{2m}{m+2} \frac{1}{\xi_1 - \eta_1} R(\xi_1, \eta_1; \xi, \eta)\right]_{\eta_1 = \xi_1 - \varepsilon} d\xi_1 + \\ &+ \frac{1}{2} \int_{\eta+\varepsilon}^{\xi} \left[\frac{\partial u(\xi_1, \eta_1)}{\partial \xi_1} - \frac{\partial u(\xi_1, \eta_1)}{\partial \eta_1}\right]_{\eta_1 = \xi_1 - \varepsilon} R(\xi_1, \xi_1 - \varepsilon;\ \xi, \eta)\, d\xi_1. \end{aligned} \tag{2.45}$$

Using the well known identity for the hypergeometric function

$$F(a, b, c, x) = \frac{\Gamma(c)\,\Gamma(c-a-b)}{\Gamma(c-a)\,\Gamma(c-b)} F(a, b, 1-c+$$

$$+a+b, 1-x) + \frac{\Gamma(c)\,\Gamma(a+b-c)}{\Gamma(a)\,\Gamma(b)} (1-x)^{c-a-b} F(c-b, c-$$

$$-a, 1+c-a-b, 1-x),$$

it is easy to see on account of (2.42) and (2.44) that

$$\left.\begin{aligned}
&\lim_{\varepsilon\to 0}\left[\left(\frac{\partial u}{\partial \xi_1} - \frac{\partial u}{\partial \eta_1}\right) R(\xi_1, \eta_1; \xi, \eta)\right]_{\eta_1=\xi_1-\varepsilon} = \\
&= \left(\frac{4}{m+2}\right)^{m/(m+2)} \frac{\Gamma\left(\frac{m+2}{2}\right)}{\Gamma^2\left(\frac{m+4}{2m+4}\right)} [(\xi_1-\eta)(\xi- \\
&\qquad -\xi_1)]^{-m/(2m+4)}\, \nu(\xi_1), \\
&\lim_{\varepsilon\to 0}\left[\frac{\partial R(\xi_1, \eta_1; \xi, \eta)}{\partial \xi_1} - \frac{\partial R(\xi_1, \eta_1; \xi, \eta)}{\partial \eta_1} - \right. \\
&\left. -\frac{2m}{m+2}\,\frac{1}{\xi_1-\eta_1} R(\xi_1, \eta_1; \xi, \eta)\right]_{\eta_1=\xi_1-\varepsilon} u(\xi_1, \xi_1-\varepsilon) = \\
&= -\frac{2\Gamma\left(\frac{m}{m+2}\right)}{\Gamma^2\left(\frac{m}{2m+4}\right)} [(\xi_1-\eta)(\xi- \\
&\qquad -\xi_1)]^{-(m-4)/(2m+4)} (\xi-\eta)^{2/(m+2)}\, \tau(\xi_1).
\end{aligned}\right\} \quad (2.46)$$

Taking into account (2.46) we get from formula (2.45) in the limit $\varepsilon \to 0$, by an obvious change of the variable of integration, the well known Darboux formula,

$$u(x, y) = \frac{\Gamma\left(\frac{m}{m+2}\right)}{\Gamma^2\left(\frac{m}{2m+4}\right)} \times$$

$$\times \int_0^1 \tau\left[x + (1-2z)\,\frac{2}{m+2}\, y^{(m+2)/2}\right] [z(1-z)]^{-(m-4)/(2m+4)}\, dz +$$

$$+ \frac{\Gamma\left(\frac{m+4}{m+2}\right)}{\Gamma^2\left(\frac{m+4}{2m+4}\right)}\, y \int_0^1 \nu\left[x + (1-2z)\,\frac{2}{m+2}\, y^{(m+2)/2}\right] [z(1-$$

$$-z)]^{-m/(2m+4)}\, dz, \quad (2.47)$$

which solves the Cauchy problem for the initial conditions given by (2.39), (2.40). The uniqueness of the solution of this problem follows from the fact that formula (2.47) is a consequence of identity (2.45) which is valid for every function that is a solution of equation (2.38) in the half-plane $\xi > \eta$ and has there continuous derivatives up to the second order. The form of the expression (2.47) shows that the found solution is stable.

We now return to equation (2.37). The Riemann function for the half-plane $\xi > \eta$ is now constructed by the method indicated in § 1 of the present chapter. Therefore we can write the identity

Not a and b of eq. (2.37)

$$u\,(\xi, \eta) = \frac{1}{2}\,u\,(\eta + \varepsilon, \eta)\,R\,(\eta + \varepsilon, \eta; \xi, \eta) +$$

$$+ \frac{1}{2}\,u\,(\xi, \xi - \varepsilon)\,R\,(\xi, \xi - \varepsilon; \xi, \eta) +$$

$$+ \frac{1}{2} \int_{\eta+\varepsilon}^{\xi} \left\{ \left[\frac{\partial u\,(\xi_1, \eta_1)}{\partial \xi_1} - \frac{\partial u\,(\xi_1, \eta_1)}{\partial \eta_1} \right] R\,(\xi_1, \eta_1; \xi, \eta) - \right.$$

$$- u\,(\xi_1, \eta_1) \left[\frac{\partial R\,(\eta_1, \eta_1; \xi, \eta)}{\partial \xi_1} - \frac{\partial R\,(\xi_1, \eta_1; \xi, \eta)}{\partial \eta_1} \right] + [2b\,(\xi_1, \eta)_r -$$

$$\left. - 2a\,(\xi_1, \eta_1)] \,u\,(\xi_1, \eta_1)\,R\,(\xi_1, \eta_1; \xi, \eta) \right\}_{\eta_1 = \xi_1 - \varepsilon} d\xi_1. \qquad (2.47a)$$

In view of the fact that the Riemann function and its derivative may have singularities of fairly high order on $\xi_1, = \eta_1$, $\xi = \xi_1$, $\eta = \eta_1$ it is, in general, impossible to take the limit $\varepsilon \to 0$, of identity (2.47a).

However if the coefficients satisfy the conditions

$$\lim_{y \to +0} y^{(1-m)/2}\,a\,(x, y) = 0 \qquad (2.48)$$

or what amounts to the same,

$$\lim_{\xi \to \eta - 0} (\xi - \eta)^{(2-m)/2+m}\,a = 0,$$

the coefficients of $\partial u/\partial \xi$ and $\partial u/\partial \eta$ in equation (2.37) will have the same singularity for $\xi = \eta$ as the corresponding coefficients of equation (2.41) and the singularity of the coefficient of $u(\xi, \eta)$ is harmless in the sense of double integration.

In this way under condition (2.48) the Riemann function for equation (2.37) behaves near the singular line $\xi = \eta$ in exactly the same way as in the case of equation (2.41) and therefore the passage to the limit $\varepsilon \to 0$ in formula (2.47a) yields the solution of the Cauchy problem for equation (2.32) satisfying the initial conditions (2.39) and (2.40).

The fact that the Cauchy problem for equation (2.32) with the initial conditions given on the curve of degeneracy may in general prove to be incorrect was first pointed out by Gellerstedt (1) and slightly later by I. S. Berezin (1). Protter (1) obtained condition (2.48) providing for the correctness of this problem in a different manner.

The example we are going to describe below shows that (2.48) is not a necessary condition for the correctness of the Cauchy problem for initial conditions given along the curve of parabolic degeneracy.

The equation

$$y^2 \frac{\partial^2 u}{\partial x^2} - \frac{\partial^2 u}{\partial y^2} + a \frac{\partial u}{\partial x} = 0, \tag{2.49}$$

where a is a real constant, is hyperbolic everywhere except on the straight line $y = 0$, and the straight line $y = 0$ is the curve of parabolic degeneracy. In the characteristic variables $\xi = x + y^2/a$, $\eta = x - y^2/a$ equation (2.49) assumes the following form

$$\frac{\partial^2 u}{\partial \xi \, \partial \eta} + \frac{a-1}{4(\xi - \eta)} \frac{\partial u}{\partial \xi} + \frac{a+1}{4(\xi - \eta)} \frac{\partial u}{\partial \eta} = 0. \tag{2.50}$$

For $\xi > \eta$ the Riemann function for equation (2.50) can be expressed by the aid of the hypergeometric function

$$R(\xi_1, \eta_1; \xi, \eta) = (\xi_1 - \eta)^{(a-1)/4} (\xi - \eta_1)^{-(a+1)/4} (\xi_1 - \eta_1)^{1/2} \times$$
$$\times F\left[\frac{1-a}{4}, \frac{1+a}{4}, 1, \frac{(\xi - \xi_1)(\eta_1 - \eta)}{(\xi - \eta_1)(\xi_1 - \eta)}\right]. \tag{2.51}$$

The form of the Riemann function (2.51) shows that for $|a| < 1$ the solution of the Cauchy problem for equation

(2.49) satisfying the boundary conditions (2.39), (2.40) can be obtained from formula (2.47) by means of a passage to the limit. The following formula is obtained after simple calculations

$$u(x, y) =$$

$$= \frac{\Gamma(1/2)}{\Gamma\left(\frac{1-a}{4}\right)\Gamma\left(\frac{1+a}{4}\right)} \int_0^1 \tau\left[x + \frac{y^2}{2}(1-2t)\right] \times$$

$$\times (1-t)^{(a-3)/4}\, t^{-(a+3)/4}\, dt +$$

$$+ y \frac{\Gamma(3/2)}{\Gamma\left(\frac{3-a}{4}\right)\Gamma\left(\frac{3+a}{4}\right)} \int_0^1 \nu\left[x + \frac{y^2}{2}(1-2t)\right](1 -$$

$$- t)^{(a-1)/4}\, t^{-(a+1)/4}\, dt.$$

If $a = -1$ we have for the Riemann function

$$R(\xi_1 \eta_1; \xi, \eta) = \sqrt{\left(\frac{\xi_1 - \eta_1}{\xi_1 - \eta}\right)},$$

and formula (2.47a) assumes the form

$$2u(\xi, \eta) = u(\eta + \varepsilon, \eta) + u(\eta, \eta - \varepsilon) +$$

$$+ \sqrt{\varepsilon}\, \frac{u(\xi, \xi - \varepsilon) - u(\eta, \eta - \varepsilon)}{\sqrt{\xi - \eta}} -$$

$$- \int_\xi^{\eta+\varepsilon} \left\{\left[\frac{\partial u(\xi_1 \eta_1)}{\partial \xi_1} - \frac{\partial u(\xi_1, \eta_1)}{\partial \eta_1}\right] \sqrt{\left(\frac{\xi_1 - \eta_1}{\xi_1 - \eta}\right)}\right\}_{\eta_1 = \xi_1 - \varepsilon} d\xi_1 -$$

$$- \frac{\sqrt{\varepsilon}}{2} \int_\xi^{\eta+\varepsilon} \frac{u(t, t-\varepsilon) - u(\eta, \eta - \varepsilon)}{(t-\eta)^{3/2}}\, dt.$$

Hence in the limit $\varepsilon \to 0$, we get

$$u(x, y) = \tau\left(x - \frac{1}{2} y^2\right) + \frac{y}{2} \int_0^1 \frac{\nu\left[x + (1-2t)\frac{y^2}{2}\right]}{\sqrt{1-t}}\, dt.$$

Also, if $a = 1$ we find similarly that the function

$$u(x, y) = \tau\left(x + \frac{1}{2} y^2\right) + \frac{y}{2} \int_0^1 \nu\left[x + (1-2t)\frac{y^2}{2}\right] \frac{dt}{\sqrt{t}}$$

is the unique solution of equation (2.49) which satisfies the conditions (2.39), (2.40).

Thus in the cases in which the coefficient of $\partial u/\partial x$ in equation (2.49) does not satisfy condition (2.48), the Cauchy problem for this equation belonging to boundary conditions given along the curve of parabolic degeneracy nevertheless has always a solution which is also unique and stable.

A very frequent occurrence of this problem is investigated in the paper of K. I. Karapatian (1) [cf. also I. L. Karol (1).]

The investigation of the Cauchy problem for equation (2.49) under increased restrictions of smoothness for the boundary conditions (2.39) and (2.40) can be found in the paper of Chi Min Iu (1).

For equation (2.33), as was shown, the curve of parabolic degeneracy $y = 0$ is itself a characteristic. It is therefore natural to expect that the Cauchy problem for this equation with the boundary conditions given on the curve of degeneracy is, in general, incorrect.

This can be easily seen from the following simple example

$$\frac{\partial^2 u}{\partial x^2} - y \frac{\partial^2 u}{\partial y^2} - \frac{1}{2} \frac{\partial u}{\partial y} = 0. \tag{2.52}$$

The general solution of this equation for the half-plane $y > 0$ is of the form

$$u(x, y) = \varphi(x + 2y^{1/2}) + \psi(x - 2y^{1/2}), \tag{2.53}$$

where φ and ψ are arbitrary twice differentiable functions.

If we demand from solution $u(x, y)$ of equation (2.52) that it satisfies condition (2.39) and in addition that the expression

$$\lim_{y \to +0} \frac{\partial u(x, y)}{\partial y} \tag{2.54}$$

remains bounded then the above solution becomes uniquely determined.

Indeed, by (2.39) and (2.54), we have from (2.53)

$$\varphi(x) + \psi(x) = \tau(x), \quad \varphi'(x) - \psi'(x) = 0$$

i.e.

$$\varphi(x) = \frac{\tau(x) + \text{const.}}{2}, \quad \psi(x) = \frac{\tau(x) - \text{const.}}{2},$$

and consequently the required solution is given by the expression

$$u(x, y) = \frac{1}{2}\tau(x + 2y^{1/2}) + \frac{1}{2}\tau(x - 2y^{1/2}).$$

Now let us replace condition (2.40) by the weaker condition

$$\lim_{y \to +0} y^{1/2} \frac{\partial u(x, y)}{\partial y} = \nu(x). \tag{2.55}$$

The solution of equation (2.52) satisfying conditions (2.39) and (2.55) is uniquely determined. It is of the following form:

$$u(x, y) = \frac{1}{2}\tau(x + 2y^{1/2}) + \frac{1}{2}\tau(x + 2y^{1/2}) + \frac{1}{2}\int_{x-2y^{1/2}}^{x+2y^{1/2}} \nu(t)\, dt.$$

§ 4. Generalizations

A series of new questions arises from the above discussion. We shall mention some of them here.

(1) As was mentioned earlier, the condition (2.48) which is sufficient for the correctness of the Cauchy problem for the case when the boundary conditions are given along the curve of parabolic degeneracy is quite restrictive.

It would be very interesting to find a condition which is weaker than (2.48) but is still sufficient for the correctness of the Cauchy problem.

A number of authors [F. I. Frenkel, (1), Conti (1), Bers (1), Haack and Hellwig (1), Hellwig (1)] have written about this problem.

(2) Let us consider the system

$$y^m k(x, y) \frac{\partial^2 u}{\partial x^2} - \frac{\partial^2 u}{\partial y^2} + a(x, y) \frac{\partial u}{\partial x} + b(x, y) \frac{\partial u}{\partial y} + \tag{2.56}$$
$$+ c(x, y) u = 0, \quad m > 0,$$

where $u = (u_1, \ldots, u_n)$ is the unknown vector, k, a, b, c are given $(u \times u)$ matrices and the matrix $k(x, y)$ is positive definite.

This system is hyperbolic for $y > 0$ and degenerates into a parabolic system for $y = 0$.

Stating the Cauchy problem in the following way its solution is far from being complete: find in the domain $y > 0$ the solution $u(x, y)$ of equation (2.56) satisfying the boundary conditions

$$\lim_{y\to+0} u(x, y) = \tau(x), \quad \lim_{y\to+0} \frac{\partial u(x, y)}{\partial y} = \nu(x), \tag{2.57}$$

where $\tau = (\tau_1, \ldots, \tau_n)$, $\nu = (\nu_1, \ldots, \nu_n)$ are given vectors.

In the case when $k(x, y)$ is the unit (diagonal) matrix, in domain $y > 0$, system (2.56) reduces, after changing the variables according to (2.36), to the form

$$\frac{\partial^2 u}{\partial\xi\,\partial\eta} + \left[\left(\frac{4}{m+2}\right)^{(m-2)/(m+2)} a\,(\xi-\eta)^{(2-m)/(2+m)} + \right.$$

$$\left. + \left(\frac{4}{m+2}\right)^{-2/(m+2)} b\,(\xi-\eta)^{2/(m+2)} - \frac{m}{2}\right] \times$$

$$\times \frac{1}{m+2}\,\frac{1}{\xi-\eta}\,\frac{\partial u}{\partial\xi} + \left[\left(\frac{4}{m+2}\right)^{(m-2)/(m+2)} a\,(\xi-\eta)^{(2-m)/(2+m)} - \right.$$

$$\left. - \left(\frac{4}{m+2}\right)^{-2/(m+2)} b\,(\xi-\eta)^{2/(m+2)} + \frac{m}{2}\right] \frac{1}{m+2}\,\frac{1}{\xi-\eta}\,\frac{\partial u}{\partial\eta} +$$

$$+ \left(\frac{4}{m+2}\right)^{2m/(m+2)} \frac{c}{4}\,\frac{1}{(\xi-\eta)^{2m/(m+2)}}\,u = 0,$$

and consequently if a, b, c are symmetrical matrices and the conditions

$$\lim_{y\to+0} y^{(1-m)/2} a_{ij}(x, y) = 0,\ i,\ j = 1, 2, \ldots, n$$

are satisfied the Cauchy problem (2.57) can be solved by means of the method in the preceding paragraph.

(3) It would be interesting in the case of equation (2.33) to clarify the conditions (i.e. the values of power m and

coefficients a, b which permit a solution of the Cauchy problem with boundary conditions (2.39) and (2.40). Also in the case when this problem is incorrect to clarify the nature of the weights $\varphi_1(x, y)$, $\varphi_2(x, y)$ for which the following problem

$$\lim_{y\to+0} \varphi_1(x, y)\, u(x, y) = \tau(x), \quad \lim_{y\to+0} \varphi_2(x, y) \frac{\partial u(x, y)}{\partial y} = \nu(x),$$

where τ and ν are given functions, is possible.

(4) For hyperbolic equations of order higher than 2 the Cauchy problem with boundary conditions on the curve of parabolic degeneracy has not yet been investigated. In the special case of the fourth order equation

$$\left(y^m \frac{\partial^2}{\partial x^2} - \frac{\partial^2}{\partial y^2}\right)\left(y^n \frac{\partial^2}{\partial x^2} - \frac{\partial^2}{\partial y^2}\right) u = 0, \; m > 0, \; n \geqslant 0$$

the problem of constructing a solution for the boundary conditions

$$\lim_{y\leftarrow+0} u(x, y) = \tau_1(x), \; \lim_{y\to+0} \frac{\partial u(x, y)}{\partial y} = \tau_2(x),$$

$$\lim_{y\to+0} \frac{\partial^2 u(x, y)}{\partial y^2} = \tau_3(x), \quad \lim_{y\to+0} \frac{\partial^3 u(x, y)}{\partial y^3} = \tau_4(x)$$

can apparently be solved by quadrature.

CHAPTER 3

THE STUDY OF THE SOLUTIONS OF SECOND ORDER ELLIPTIC EQUATIONS FOR A DOMAIN, THE BOUNDARY OF WHICH INCLUDES A SEGMENT OF THE CURVE OF PARABOLIC DEGENERACY

§ 1. The Linear Elliptic Partial Differential Equation of the Second Order

Let us consider the linear elliptic partial differential equation of second order in two independent variables in its canonical form

$$L(u) = \frac{\partial^2 u}{\partial x^2} + \frac{\partial^2 u}{\partial y^2} + A(x, y)\frac{\partial u}{\partial x} + B(x, y)\frac{\partial u}{\partial y} + C(x, y)\, u = 0, \tag{3.1}$$

the coefficients of which are real analytic functions given in some singly-connected domain D_0.

The solution $u(x, y)$ of equation (3.1) is said to be regular if it is continuous together with its derivatives up to and including those of second order. It is known that a solution $u(x, y)$, of equation (3.1) which is regular in domain D_0 is an analytic function of the variables x, y.

We give here an account of some results from the theory of equation (3.1) due to I. N. Vekua (1).

The coefficients A, B, C and the regular solution $u(x, y)$ of equation (3.1) can be continued analytically for complex values of x, y, and therefore, introducing the new independent variables $z = x + iy$, $\zeta = x - iy$ this equation can be written for some cylindrical domain (D, D^*) of the variables $z\ D$, $\zeta \in D^*$ in the following form:

$$L(u) = \frac{\partial^2 u}{\partial z\, \partial \zeta} + a \frac{\partial u}{\partial z} + b \frac{\partial u}{\partial \zeta} + cu = 0, \tag{3.2}$$

where $4\, a(z, \zeta) = A + iB$, $4\, b(z, \zeta) = A - iB$, $4\, c(z, \zeta) = C$,

$$2 \frac{\partial}{\partial z} = \frac{\partial}{\partial x} - i \frac{\partial}{\partial y}, \quad 2 \frac{\partial}{\partial \zeta} = \frac{\partial}{\partial x} + i \frac{\partial}{\partial y}.$$

Following Vekua we note that it is possible to introduce the complex Riemann function $R(z, \zeta; z_1, \zeta_1)$, in the same way as it was done in § 1 of Chapter 2 which in terms of the variables is the solution $z \in D$, $\zeta \in D^*$ of the adjoint equation of (3.2)

$$M(v) = \frac{\partial^2 v}{\partial z \partial \zeta} - \frac{\partial}{\partial z}(av) - \frac{\partial}{\partial \zeta}(bv) + cv = 0, \tag{3.3}$$

which satisfies the conditions

$$\left.\begin{aligned} &R(z_1, \zeta; z_1, \zeta_1) = \exp\left[\int_{\zeta_1}^{\zeta} a(z_1, \zeta_2)\, d\zeta_2\right], \\ &R(z, \zeta_1; z_1, \zeta_1) = \exp\left[\int_{z_1}^{z} b(z_2, \zeta_1)\, dz_2\right], \\ &R(z, \zeta; z, \zeta) = 1. \end{aligned}\right\} \tag{3.4}$$

Regarded as a function of $z_1 \in D$, $\zeta_1 \in D^*$, $R(z, \zeta; z, \zeta) \neq 1$ satisfies equation (3.2).

In terms of this Riemann function we can prove directly the formula which is analogous to (2.13):

$$\begin{aligned} u(z, \zeta) = {} & R(z, \zeta_0; z, \zeta)\, \varphi(z) + R(z_0, \zeta; z, \zeta)\, \psi(\zeta) - \\ & - R(z_0, \zeta_0; z, \zeta)\, \varphi(z_0) + \\ & + \int_{z_0}^{z} \left[b(t, \zeta_0)\, R(t, \zeta_0; z, \zeta) - \frac{\partial R(t, \zeta_0; z, \zeta)}{\partial t} \right] \varphi(t)\, dt + \\ & + \int_{\zeta_0}^{\zeta} \left[a(z_0, \tau)\, R(z_0, \tau; z, \zeta) - \frac{\partial R(z_0, \tau; z, \zeta)}{\partial \tau} \right] \psi(\tau)\, d\tau, \end{aligned} \tag{3.5}$$

where $\varphi(z) = u(z, \zeta_0)$, $\psi(\zeta) = u(z_0, \zeta)$ are arbitrary holomorphic functions in variables z and ζ, $z_0 = x_0 + iy_0$, $\zeta_0 = x_0 - iy_0$ — are fixed points contained in domains D and D^* respectively.

Formula (3.5) furnishes a general complex representation for all analytic solutions of equation (3.2) for $z \in D$, $\zeta \in D^*$.

In the case when the variables ζ and ζ_0 are the complex conjugates of z and z_0, i. e. when $\zeta = \bar{z}$, $\zeta_0 = \bar{z}_0$, the Riemann function $R(z, \bar{z}; z_0, \bar{z}_0)$ assumes real values only and formula (3.5) furnishes the general representation of all real regular solutions of equation (3.1):

$$u(x,y) = \mathrm{Re}\left\{R(z, \bar{z}_0; z, \bar{z})\,\varphi(z) + R(z_0, \bar{z}_0; z, \bar{z})\,\varphi(z_0) + \right.$$
$$\left. + \int_{z_0}^{z}\left[b(t, \bar{z}_0)\,R(t, \bar{z}_0; z, \bar{z}) - \frac{\partial R(t, \bar{z}_0; z, \bar{z})}{\partial t}\right]\varphi(t)\,dt\right\}. \tag{3.6}$$

If A, B, C are integral functions of variables x, y then representation (3.6) is valid for every finite singly connected domain of x, y.

Since the Riemann function as a function of its last pair of arguments is a solution of equation (3.1) and also identities (3.4) are true, each of the expressions

$$\mathrm{Re}\,[R(z_0, \bar{z}_0; z, \bar{z})\,\varphi(z_0)],\quad \mathrm{Re}\int_{z_0}^{z} b(t, \bar{z}_0)\,R(t, \bar{z}_0; z, \bar{z})\,\varphi(t)\,dt,$$

is a solution of equation (3.1).

Consequently the function

$$u_1(x,y) = \mathrm{Re}\left\{R(z, \bar{z}_0; z, \bar{z})\,\varphi(z) - \right.$$
$$\left. - \int_0^1 \frac{\partial}{\partial\sigma} R[z_0 + (z - z_0)\,\sigma, \bar{z}_0; z, \bar{z}]\cdot\varphi[z_0 + (z - z_0)\,\sigma]\,d\sigma\right\} \tag{3.7}$$

is also a solution of equation (3.1).

In the same way we conclude that the function

$$v_1(x,y) = \mathrm{Re}\left\{R(z, \bar{z}; z, \bar{z}_0)\,\varphi(z) - \right.$$
$$- \int_0^1 \frac{\partial}{\partial\sigma} R[z, \bar{z}; z_0 + (z - z_0)\,\sigma, \bar{z}_0]\cdot\varphi[z_0 + (z - z_0)\,\sigma]\,d\sigma$$

is a solution of the equation adjoint to (3.1).

$$M(v) = \frac{\partial^2 v}{\partial x^2} + \frac{\partial^2 x}{\partial y^2} - \frac{\partial}{\partial x}(Av) - \frac{\partial}{\partial y}(Bv) + Cv = 0. \quad (3.8)$$

Substituting in (3.7) the function $k \ln (z - z_0)$, where k is an arbitrary real constant, for $\varphi(z)$ we get the so-called fundamental solution of equation (3.1) with a logarithmic singularity at point $z = z_0$:

$$\omega_1(z, z_0) = k[g(y, z_0)] \ln |z - z_0| - g_0(z, z_0),$$

where

$$g(z, z_0) = R(z_0, \bar{z}_0; z, \bar{z}),$$

$$g_0(z, z_0) = \mathrm{Re} \int_0^1 \ln \sigma \cdot \frac{\partial}{\partial \sigma} R[z_0 + (z - z_0)\sigma, \bar{z}_0; z, \bar{z}]\, d\sigma.$$

The solution of the adjoint equation (3.8) can be obtained in similar manner:

$$\omega_1^*(z, z_0) = k[g(z_0, z)] \ln |z - z_0| - g_0^*(z, z_0),$$

where

$$g_0^* = \mathrm{Re} \int_0^1 \ln \sigma \cdot \frac{\partial}{\partial \sigma} R[z, \bar{z}; z_0 + (z - z_0)\sigma, \bar{z}_0]\, d\sigma.$$

It is clear that the expression

$$\omega(z, z_0) = \omega_1(z, z_0) - \int_{z_1}^{z_0} d\xi \int_{\bar{z}_1}^{\bar{z}_0} R(z_0, \bar{z}_0; \xi, \bar{\xi}) M_{\xi s}[\omega_1(z, \bar{\xi})]\, d\xi, \quad (3.9)$$

where $z_1 = x_1 + iy_1$ is a fixed point regarded as a function of the variables x, y, is a fundamental solution of equation (3.1) and as a function of variables x_0, y_0 if a fundamental solution of the adjoint equation (3.8) with logarithmic singularity at $z = z_0$.

In what follows we shall assume that $k = -1/2\pi$, and the function (3.9) will then be said to be the normalized fundamental solution. The existence of the normalized fundamental solution can be proved by other methods [cf. Lèvy (1)].

Let D be a singly-connected domain contained within the domain D_0, and having a simple Jordan curve Γ as its boundary.

The following Dirichlet problem, or as it is commonly called the first boundary problem in connection with equation (3.1) is well investigated: find the solution $u(x, y)$ equation (3.1) which is regular in domain D, continuous in the closed domain $\bar{D}$ and satisfies the boundary condition

$$u|_{\Gamma} = f, \tag{3.10}$$

where f is a given continuous function.

The following two statements are familiar from the theory of elliptic equations: (a) in a sufficiently small domain the Dirichlet problem for equation (3.1) has always a unique solution, (b) if the homogeneous Dirichlet problem ($f = 0$) has only a trivial solution, then the non-homogeneous Dirichlet problem can always be solved.

Alternative (b) indicates the importance of finding conditions for the homogeneous Dirichlet problem to have only the trivial solution. One of the classical conditions for this is the non-positivity of coefficient C in the considered domain.

Thus if

$$C \leqslant 0 \tag{3.11}$$

the Dirichlet problem (3.10) always has a unique solution.

Assuming the smoothness of contour Γ and the continuity of solutions $u(x, y)$ and $v(x, y)$ of equations (3.1) and (3.8) respectively, together with that of their partial derivatives of first order in domain $\bar{D}$, we have the following classical formula

$$\int_{\Gamma} \left[u \frac{\partial v}{\partial N} - v \frac{\partial u}{\partial N} - uv\big(a \cos(N, x_0) + b \cos(N, y_0)\big)\right] ds = 0, \tag{3.12}$$

where N is the inner normal of Γ, and s is the length of the curve measured from a fixed point in the positive direction.

Denote by $G(x, y; x_0, y_0)$ the function

$$\omega(z, z_0) - \omega_0(z, z_0), z, z_0 \in D,$$

where $\omega(z, z_0)$ is the normalized fundamental solution (3.9) and $\omega_0(z, z_0)$ is a solution with respect to x, y, of equation (3.8) which is regular in domain D, is continuous in domain $\bar{D}$ and satisfies the condition

$$\omega_0(z, z_0) = \omega(z, z_0), z_0 \in \Gamma. \tag{3.13}$$

The function $G(x, y; x_0, y_0)$ is called the Green's function for the Dirichlet problem.

If condition (3.11) is satisfied, the Green's function obviously exists also as a function of variables x, y; it is a solution of equation (3.1).

Separating the point z from the domain D by a circle of sufficiently small radius ε, and applying formula (3.12) in the remaining part of D with $v(x_0, y_0) = G(x, y; x_0, y_0)$, we get, on passing to the limit $\varepsilon \to 0$ and noting (3.13):

$$u(x, y) = \int_{\Gamma} \frac{\partial G[z, z_0(s)]}{\partial N_{z_0}} f(s)\, ds. \tag{3.14}$$

The representation (3.14) of the solution of the Dirichlet problem for equation (3.1) is valid also in those cases when contour Γ has corners and the coefficients [cf. Vekua (2)] a, b, c satisfy much weaker restrictions than analyticity.

§ 2. Elliptic Systems of Second Order

Consider the following system of linear partial differential equations of the second order

$$A \frac{\partial^2 u}{\partial x^2} + 2B \frac{\partial^2 u}{\partial x \partial y} + C \frac{\partial^2 u}{\partial y^2} + A_1 \frac{\partial u}{\partial x} + \\ + B_1 \frac{\partial u}{\partial y} + C_1 u = 0, \tag{3.15}$$

where A, B, C, A_1, B_1, C_1 are real $(n \times n)$ matrices whose elements are defined in some domain D_0 as functions of x, y and $u = (u_1, \ldots, u_n)$ is the solution vector to be found.

According to the general definition (cf. Chapter 1, § 4) the system (3.15) is called elliptic in domain D_0, if at every point of this domain the roots of the characteristic equation

$$\det|A + 2B\lambda + C\lambda^2| = 0 \tag{3.16}$$

are all complex.

From the point of view of boundary problems (especially from the point of view of the Dirichlet problem) the above definition of ellipticity has different meanings when applied to one second order equation or to a system of second order equations.

In the preceding paragraph we saw that for a second order equation of the elliptic type the Dirichlet problem is of the Fredholm type. More precisely, for one second order elliptic equation with sufficiently smooth coefficients interpreted in a finite domain, having a sufficiently well-behaved boundary, the homogeneous Dirichlet problem has a finite number of linearly independent solutions. The non-homogeneous Dirichlet problem can be solved only if certain additional conditions are satisfied, the number of which coincides with the number of linearly independent solutions for the homogeneous problem. Specifically, the Dirichlet problem has always a solution for a sufficiently small domain and this solution is unique.

For the elliptic system (3.15) the Dirichlet problem can be formulated in the following way: Find in domain $D \subset D_0$ the boundary with Γ, the regular solution $u = (u_1, \ldots, u_n)$ of system (3.15) which is continuous in the closed region $\overline{D}$ and satisfies the boundary condition

$$u|_\Gamma = f, \tag{3.17}$$

where f is a continuous vector given in Γ.

For system (3.15) the Dirichlet problem is not, in general, of Fredholm type. This can be seen easily from simple examples [cf. A. V. Bitsadze (1, 2)].

In fact, the system

$$\begin{aligned} \frac{\partial^2 u_1}{\partial x^2} - \frac{\partial^2 u_1}{\partial y^2} - 2\frac{\partial^2 u_2}{\partial x\,\partial y} &= 0, \\ 2\frac{\partial^2 u_1}{\partial x\,\partial y} + \frac{\partial^2 u_2}{\partial x^2} - \frac{\partial^2 u_2}{\partial y^2} &= 0, \end{aligned} \tag{3.18}$$

is elliptic according to the definition given above since the characteristic equation (3.16) is now

$$(\lambda^2 + 1)^2 = 0, \tag{3.19}$$

which has only complex roots.

Using the notation $u_1 + iu_2 = w$, $z = x + iy$, $\bar{z} = x - iy$ system (3.18) can be written in the form

$$\frac{\partial^2 w}{\partial \bar{z}^2} = 0, \tag{3.20}$$

where

$$2\frac{\partial}{\partial \bar{z}} = \frac{\partial}{\partial x} + i\frac{\partial}{\partial y}.$$

From (3.20) we conclude immediately that the regular solutions of (3.18) can be represented in the form

$$u_1 + iu_2 = \bar{z}\varphi(z) + \psi(z), \tag{3.21}$$

where $\varphi(z)$ and $\psi(z)$ are arbitrary holomorphic functions of the variable z.

Let D be a circle with radius R and centre at point z_0.

All solutions of system (3.18) which are regular inside the domain D and vanish on its boundary Γ: $(z - z_0)(\bar{z} - \bar{z}_0) = R^2$, are given by the formula

$$u_1 + iu_2 = \left[1 - \frac{(z - z_0)(\bar{z} - \bar{z}_0)}{R^2}\right]\psi(z),$$

where $\psi(z)$ is an arbitrary holomorphic function in domain D. Consequently the homogeneous Dirichlet problem for circle D has an infinite set of linearly independent solutions:

$$u_{1n} = \left(1 - \frac{r^2}{R^2}\right) r^n (\alpha_n \cos n\varphi - \beta_n \sin n\varphi),$$

$$u_{2n} = \left(1 - \frac{r^2}{R^2}\right) r^n (\alpha_n \sin u\varphi + \beta_n \cos n\varphi),$$

$$n = 0, \ 1, \ldots, \quad z - z_0 = re^{i\varphi}.$$

We should not think that the multiplicity of the roots of characteristic equation (3.19) is the reason why the Dirichlet problem is not of the Fredholm type. In fact the system

$$\begin{aligned}\frac{\partial^2 u_1}{\partial x^2} - \frac{\partial^2 u_1}{\partial y^2} + \sqrt{2}\,\frac{\partial^2 n_2}{\partial x\,\partial y} &= 0,\\ \frac{\partial^2 u_2}{\partial x^2} - \frac{\partial^2 u_2}{\partial y^2} - \sqrt{2}\,\frac{\partial^2 u_1}{\partial x\,\partial y} &= 0\end{aligned} \qquad (3.22)$$

is elliptic since its characteristic equation $\lambda^4 + 1 = 0$ has only complex roots. These roots are all simple, nevertheless the Dirichlet problem for this system in circle D has an infinite set of linearly independent solutions

$$u_1^{(1)} = \alpha_1 (1 - r^2), \quad u_2^{(1)} = \beta_1 (1 - r^2);$$

$$u_1^{(2)} = 2\sqrt{2}\alpha_2 (x - x_0)(y - y_0)(1 - r^2) + $$
$$+ \beta_2 [(x - x_0)^2 - (y - y_0)^2](1 - r^2);$$

$$u_1^{(2)} = \alpha_2 [(x - x_0)^2 - (y - y_0)^2](1 - r^2) + $$
$$+ 2\sqrt{2}\beta_2 (x - x_0)(y - y_0)(1 - r^2);$$

where $r^2 = (x - x_0)^2 + (y - y_0)^2/R^2$ and $\alpha_1, \beta_1, \alpha_2, \beta_2, \ldots$ are arbitrary real numbers. This is easily seen by noting that all regular solutions of system (3.22) can be represented in the form

$$u_1 + iu_2 = \varphi(\lambda z - \bar{z}) + \psi(\varkappa z + \bar{z}),$$

where φ and ψ are arbitrary holomorphic functions and $\lambda = i/(1 + \sqrt{2})$

If the coefficients of the main part of system (3.15) satisfy in domain D_0 the condition for the positive definiteness of a quadratic form

$$\eta A \eta + \eta B \xi + \xi B \eta + \xi C \xi \geqslant 0, \qquad (3.23)$$

where $\eta = (\eta_1, \ldots, \eta_n)$, $\xi = (\xi_1, \ldots, \xi_n)$ (and it is also assumed that equality takes place only for $\eta = \xi = 0$), then as was pointed out in a number of special cases already by Somigliana (1) the Dirichlet problem for system (3.15) becomes of the Fredholm type.

This can be proved by means of different methods.

We can usefully develop for example, the method of generalized potentials, whereas the fundamental solutions can be obtained by the method of Lèvy, at least for small domains [cf. also M. I. Vishik (1)].

If the Somigliana condition (3.23) or as it is also called, the condition for strong ellipticity is satisfied, the following Fredholm alternative arises: the inhomogeneous Dirichlet problem can always be solved if the homogeneous problem ($f = 0$) has only trivial solution.

For a system of the form

$$L(u) = \frac{\partial}{\partial x}\left(A\frac{\partial u}{\partial x} + B\frac{\partial u}{\partial y}\right) + \frac{\partial}{\partial y}\left(B\frac{\partial u}{\partial x} + C\frac{\partial u}{\partial y}\right) + \\ + A_1\frac{\partial u}{\partial x} + B_1\frac{\partial u}{\partial y} + C_1 u = 0, \tag{3.24}$$

if matrices A_1 and B_1 are symmetric and if condition (3.23) is satisfied, a sufficient condition for the Dirichlet problem to have only trivial solution consists in the positive definiteness of the quadratic form

$$\eta\left(\frac{\partial A_1}{\partial x} + \frac{\partial B_1}{\partial y} - 2C_1\right)\eta \geqslant 0 \quad (x, y) \in D_0. \tag{3.25}$$

Indeed, if we apply Green's formula to domain D (it is assumed that the contour Γ, the coefficients of system (3.24) and the solution vector $u(x, y)$ satisfy the condition necessary for the applicability of this formula) and take into account that $u(x, y)$ is a solution of the homogeneous Dirichlet problem, we find after integration from the formula

$$\frac{\partial}{\partial x}\left(uA\frac{\partial u}{\partial x} + uB\frac{\partial u}{\partial y} + \frac{1}{2}uA_1 u\right) + \frac{\partial}{\partial y}\left(uB\frac{\partial u}{\partial x} + \right.$$
$$\left. + uC\frac{\partial u}{\partial y} + \frac{1}{2}uB_1 u\right) = uL(u) + \frac{\partial u}{\partial x}A\frac{\partial u}{\partial x} +$$
$$+ \frac{\partial u}{\partial x}B\frac{\partial u}{\partial y} + \frac{\partial u}{\partial y}B\frac{\partial u}{\partial x} + \frac{\partial u}{\partial y}C\frac{\partial u}{\partial y} +$$
$$+ \frac{1}{2}u\left(\frac{\partial A_1}{\partial x} + \frac{\partial B_1}{\partial y} - 2C_1\right)u$$

that

$$\iint_{\Gamma}\Big[\frac{\partial u}{\partial x} A \frac{\partial u}{\partial x} + \frac{\partial u}{\partial x} B \frac{\partial u}{\partial y} + \frac{\partial u}{\partial y} B \frac{\partial u}{\partial x} + \frac{\partial u}{\partial y} C \frac{\partial u}{\partial y} +$$

$$+ \frac{1}{2} u \left(\frac{\partial A_1}{\partial x} + \frac{\partial B_1}{\partial y} - 2C_1\right) u \Big] \mathrm{d}x\,\mathrm{d}y = 0, \tag{3.26}$$

whence, by (3.23) and (3.25) it can be concluded that $u(x, y) = 0$ everywhere in domain D.

With $A_1 = \text{const.}$, $B_1 = \text{const.}$, condition (3.25) assumes the form

$$\eta C_1 \eta \leqslant 0. \tag{3.27}$$

Note that if

$$A = \alpha(x, y) E, \quad B = 0, \quad C = \beta(x, y) E, \quad A_1 = \alpha_1(x, y) E,$$

$$B_1 = \beta_1(x, y) E,$$

where E is the unit (diagonal) matrix, and α, β, α_1, β_1 are given scalar functions such that $\alpha > 0$, $\beta > 0$, then the positive definiteness of matrix $C_1(x, y)$, implies that the length $R = (\Sigma u_i^2)^{1/2}$ of the solution vector $u(x, y)$ of system (3.24) cannot attain its maximum in the domain D.

Indeed, suppose that $R(x, y)$ has a maximum at an interior point (x, y) of domain D. On the one hand, we must have at this point

$$\frac{\partial R}{\partial x} = \frac{\partial R}{\partial y} = 0,$$

$$\alpha \frac{\partial^2 R}{\partial x^2} + \beta \frac{\partial^2 R}{\partial y^2} \leqslant 0,$$

and on the other hand by (3.24) and (3.26) and the positive definiteness of matrix $C_1(x, y)$, we have

$$\alpha \frac{\partial^2 R}{\partial x^2} + \beta \frac{\partial^2 R}{\partial y^2} = \left[- u C_1 u + \alpha \left(\frac{\partial u}{\partial x}\right)^2 + \beta \left(\frac{\partial u}{\partial y}\right)^2\right] \frac{1}{R} > 0$$

which contradicts the condition (3.27).

The simple example of an elliptic system

$$\frac{\partial^2 u_1}{\partial x^2} - \frac{\partial^2 u_1}{\partial y^2} - 2\frac{\partial^2 u_2}{\partial y^2} = 0$$
$$2\frac{\partial^2 u_1}{\partial x^2} + \frac{\partial^2 u^2}{\partial x^2} - \frac{\partial^2 u_2}{\partial y^2} = 0 \qquad (3.28)$$

shows that although the condition for strong ellipticity is not satisfied, since in this case

$$\eta A \eta + \eta B \xi + \xi B \eta + \xi C \xi = (\eta_1 + \eta_2)^2 - (\xi_1 + \xi_2)^2,$$

nevertheless the Dirichlet problem has always a unique solution for this system.

This can be easily seen if we note that the regular solutions of system (3.28) can be represented by the formula

$$u_1 = \mathrm{Re}\,[\bar{z}\varphi'(z) + \psi(z)],$$
$$u_2 = \mathrm{Re}\,[-\bar{z}\varphi'(z) - 2\varphi(z) - \psi(z)],$$

where $\varphi(z)$ and $\psi(z)$ are arbitrary holomorphic functions. We return now to system (3.18). At first glance it may seem strange that for this system the boundary problem

$$u_1|_\Gamma = f_1, \quad \left[\frac{\partial u_1}{\partial x} - \frac{\partial u_2}{\partial y}\right]_\Gamma = f_2$$

is always possible, the homogeneous problem permitting the solution $u_1 = 0$, $u_2 = ax + b$. For this system the Dirichlet problem

$$(u_1 + iu_2)_\Gamma = f,$$

is in general unsolvable. For example if D is the unit circle $|z| < 1$, then for the solvability of the Dirichlet problem it is necessary that the function $tf(t)$ be the boundary value of a function which is holomorphic inside domain D i.e. that the conditions

$$\int_\Gamma t^k f(t)\, dt = 0, k = 1, 2 \ldots \qquad (3.29)$$

be satisfied. When this is the case the Dirichlet problem is always solvable and the solution itself has the form

$$u_1 + iu_2 = (1 - z\bar{z})\,\psi(z) + \frac{\bar{z}}{2\pi i}\int \frac{tf(t)\,dt}{t - z},$$

where $\psi(z)$ is an arbitrary holomorphic function.

Let us now consider the elliptic system

$$A\frac{\partial^2 u}{\partial x^2} + 2B\frac{\partial^2 u}{\partial x\,\partial y} + C\frac{\partial^2 u}{\partial y^2} = 0 \qquad (3.30)$$

with constant coefficients. Let $\lambda_1, \ldots, \lambda_\mu; \lambda_1 \ldots, \lambda_\mu$ denote the roots of the characteristic equation of (3.16) and $k_1, \ldots, k_\mu$ their respective multiplicity.

All regular solutions of system (3.30) can be represented in a singly-connected region by the formula

$$u = \mathrm{Re}\sum_{j=1}^{\mu}\sum_{l=1}^{k_j}\sum_{m=0}^{l-1} C_{lm}^{(j)}\bar{z}_j^m\,\varphi_{jl}^{(m)}(z_j), \qquad (3.31)$$

where φ_{jl} are arbitrary holomorphic functions of the respective complex arguments $z_j = x + \lambda_j y$, the upper index m of function φ_{jl} indicates the order of the derivative with respect to z_j, and $C_{lm}^{(j)}$ are constant vectors which can be expressed in terms of the coefficients of system (3.30) alone. To find them we have only to solve a system of linear algebraic equations.

If $C_{lm}^{(j)}$ are real we shall call system (3.30) weakly connected whenever the system comprising the n vectors $\{C_{l0}^{(j)}\}$ is linearly independent. For the weakly connected system (3.30) the Dirichlet problem is obviously correct. If $C_{lm}^{(j)}$ are complex, conditions that are analogous to weak connexion can be found which will make the Dirichlet problem for the system (3.30) Fredholm type. It is interesting to find criteria for the weak connexion of system (3.30) in terms of the coefficients of this system.

Formula (3.31) makes it possible to reduce the investigation of any linear boundary problem (Dirichlet, Poincaré etc.) to the equivalent boundary problem in the theory of holomorphic functions. Hence it is clear that we cannot in general expect the appearance of the Fredholm type alternatives for these boundary problems. For elliptic systems not satisfying the strong ellipticity condition (3.23), the alternatives for the boundary problems are apparently strongly related to the well known alternatives occurring in the theory of general linear equations in abstract metric spaces [about these alternatives, cf. Hausdorff (1), C. M. Nikolskij (1)].

The special elliptic system

$$\Delta u + A\frac{\partial u}{\partial x} + B\frac{\partial u}{\partial y} + Cu = 0, \tag{3.32}$$

where Δ is the Laplace operator, A, B, C are real quadratic matrices of order n, obviously satisfies the Somigliani condition (3.23).

In the case when A, B, C are analytic matrix functions of the variables x, y, writing (3.32) in complex form

$$\frac{\partial^2 u}{\partial z\,\partial \zeta} + a\frac{\partial u}{\partial z} + b\frac{\partial u}{\partial \zeta} + cu = 0,$$

$$z = x + iy \in D, \quad \zeta = x - iy \in D^*,$$

in the same way as in § 2 of Ch. 2, we can introduce the complex Riemann matrix function $R(z, \zeta; z_1, \zeta_1)$ and write the general complex representation of all regular solutions of system (3.32) in the form

$$u(x, y) = \mathrm{Re}\Big\{ R(z, \bar{z}_0; z, \bar{z})\,\varphi(z) + R(z_0, \bar{z}_0; z, \bar{z})\,\varphi(z_0) + \\ + \int_{z_0}^{z}\Big[b(t, \bar{z}_0)\,R(t, \bar{z}_0; z, \bar{z}) - \frac{\partial R(t, \bar{z}_0; z, \bar{z})}{\partial t}\Big]\varphi(t)\,dt, \tag{3.33}$$

where $\varphi(z) = \{\varphi_1(z), \ldots, \varphi_n(z)\}$ in an arbitrary holomorphic vector.

As in the preceding paragraph, formula (3.33) furnishes the square matrix $\omega(z, z_0) = \omega(x, y; x_0, y_0)$ which is of order n, and which as a function of x_0, y_0, is the fundamental solution of system (3.32) and as a function of x, y is the fundamental solution of the adjoint system of (3.32)

$$\Delta v - \frac{\partial}{\partial x}(vA) - \frac{\partial}{\partial y}(vB) + vC = 0$$

with a logarithmic singularity at point $z = z_0$.

The Dirichlet problem (3.17) has been investigated for the system (3.32) with the same completeness as for the single equation (1.1) [cf. I. N. Vekua (1), A. V. Bitsadze (3)]. It is found that if

$$\eta\left(\frac{\partial A}{\partial x} + \frac{\partial B}{\partial y} - 2C\right)\eta \geqslant 0. \qquad (3.34)$$

Then the Dirichlet problem for system (3.32) has always the unique solution

$$u(x, y) = \int \frac{\partial G[z, z_0(s)]}{\partial N_{z_0}} f(s)\, ds, \qquad (3.35)$$

where $G(z, z_0)$ is the matrix Green's function; which always exists whenever the contour Γ is sufficiently smooth.

§ 3. The Dirichlet Problem for Second Order Elliptic Equations in a Domain, the Boundary of which includes a Segment of the Curve of Parabolic Degeneracy

Consider the equations

$$L(u) = y^m \frac{\partial^2 u}{\partial x^2} + \frac{\partial^2 u}{\partial y^2} + a\frac{\partial u}{\partial x} + b\frac{\partial u}{\partial y} + cu = 0 \qquad (3.36)$$

and

$$L(u) = \frac{\partial^2 u}{\partial x^2} + y^m \frac{\partial^2 u}{\partial y^2} + a\frac{\partial u}{\partial x} + b\frac{\partial u}{\partial y} + cu = 0. \qquad (3.37)$$

The coefficients $a(x, y)$, $b(x, y)$, $c(x, y)$ are real functions given in some domain D_0 containing a segment of the straight line $y = 0$. Let m be a positive real number.

In domain D_0 both equations (3.36) and (3.37) are elliptic for $y > 0$ and parabolic for $y = 0$.

The portion of domain D_0 in which $y > 0$, will be called the elliptic part.

The application of the non-singular transformation

$$\xi = x, \ \eta = \frac{2}{m+2} y^{(m+2)/2}$$

reduces equation (3.36) in domain D_0 to

$$\frac{\partial^2 u}{\partial \xi^2} + \frac{\partial^2 u}{\partial \eta^2} + A \frac{\partial u}{\partial \xi} + B \frac{\partial u}{\partial \eta} + Cu = 0.$$

Equation (3.37) reduces also to the same form by the application of the transformation

$$\xi = x, \ \eta = \frac{2}{2-m} y^{(2-2)/2} \text{ for } m \neq 2,$$

$$\eta = \ln y \text{ for } m = 2.$$

In what follows it will be assumed that in the elliptic part of domain D_0 the coefficients a, b and c of equations (3.36), (3.37) are real analytic functions and that $c(x, y)$ satisfies condition (3.11).

Let D be a singly-connected region with the piecewise smooth contour Γ, contained inside the elliptic part of domain D_0.

Repeating the considerations explained in § 1 of the present chapter we conclude that the Dirichlet problem for equation (3.36) and also for equation (3.37) with a continuous boundary condition f on Γ, always possesses a solution $u(x, y)$, which is unique and can be represented in the form of an integral

$$u(x, y) = \int_\Gamma k(x, y, s) f(s) \, ds. \tag{3.38}$$

The kernel of this representation is an analytic solution in the variables x, y, and it depends solely on the domain D.

Assume now that D is a singly-connected domain the boundary of which consists of two parts $\Gamma = \sigma + AB$, where σ is a smooth Jordan curve with its endpoints at the points $A(0, 0)$,

B (1, 0), lying inside the elliptic part of D_0 and AB is a segment of the line $y = 0$.

We note the following obvious but extremely important property of the operator L, on the left side of equation (3.36) (or equation (3.37)): If $v(x, y)$ is a continuous function whose partial derivatives up to the second are also continuous, and if everywhere in domain D, $L(v) < 0$ then $v(x, y)$ cannot attain a negative minimum; if conversely everywhere in domain D, $L(v) > 0$ then inside this domain $v(x, y)$ cannot attain a positive maximum.

In this paragraph the Dirichlet problem will be studied in the following form: Let $f(x, y)$ be continuous in the closed domain $\bar{D}$. It is required to find in domain $\bar{D}$ the regular solutions $u(x, y)$ of equation (3.36) [or of equation (3.37)] which coincide on the boundary $\Gamma = \sigma + AB$ with the function f.

By (3.11) the uniqueness of the solution for this Dirichlet problem is obvious.

It will be seen below that from the point of view of the existence of a solution for the Dirichlet problem equations (3.36) and (3.37) are not always identical.

Denote by D_h the domain that contains those points of D which satisfy the condition $h > y$ where h is a sufficiently small positive number, (Fig. 4). We know already that equation (3.36) [or equation (3.37)] has a unique solution which is regular in domain D_h and assumes the values of function f on its boundary. By condition (3.11) we have $|u_h(x, y)| \leq M$, in the closed domain D_h where $M = \max |f(x, y)|$ is calculated for the whole of the closed domain D. Denote by h^* an arbitrary fixed value of h. Starting from h^* the families $\{u_h(x, y)\}$ are all completely determined and equivalent for every $h < h^*$. On the other hand, by (3.38) we have the representation

$$u_h(x, y) = \int_{\Gamma_*} k^*(x, y, s)\, u_h(s)\, ds, \tag{3.39}$$

where Γ_* is the boundary of $D_h{}^*$. Hence the uniform continuity of the family $\{u_h(x, y)\}$ follows immediately.

According to the well known theorem of Arzela, we can select a uniformly convergent sequence from the family $\{u_h\}$ the limit of which $u(x, y) = \lim u_{h_i}(x, y)$ is by (3.39), a regular solution of equation (3.36) or equation (3.51) in the domain D, and coincides with the function f on the segment σ.

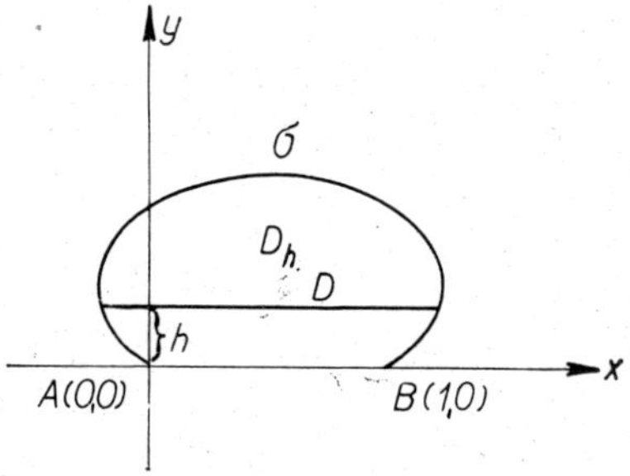

FIG. 4

The following question has still to be answered: does the function $u(x, y)$ assume the values of f on the segment AB? In the case of equation (3.36) this question can always be answered in the affirmative, whereas in the case of equation (3.37) the answer is not always in the affirmative.

Consider first the case of equation (3.36).

We show beforehand that for every point $Q(x_0, 0)$, $0 < x_0 < 1$ there exists a function $v(x, y)$, the so-called barrier, possessing the following properties: (a) it is continuous in the semicircular neighbourhood ω_{x0}:

$$(x - x_0)^2 + y^2 < \varrho^2, \; y \geqslant 0,$$

of point Q; (b) it vanishes at Q, (c) it is positive for every other point of the neighbourhood ω_{x0}; (d) everywhere in this neighbourhood

$$L(v) < 0. \qquad (3.40)$$

In the case considered we may take the function

$$v(x, y) = (x - x_0)^2 + y^\beta, \qquad (3.41)$$

where

$$0 < \beta < 1. \tag{3.42}$$

as the barrier.

Indeed, it is obvious that function $v(x, y)$ defined by (3.41) satisfies (a), (b), (c). Condition (d) can also be easily checked. In fact, we get by direct calculation

$$L(v) = 2y^m + \beta(\beta - 1)y^{\beta-2} + 2a(x - x_0) + \beta b y^{\beta-1} + cv, \tag{3.43}$$

whence it follows at once that for sufficiently small y the sign of $L(v)$ is the same as that of $\beta(\beta - 1)$ and therefore, by (3.42) there exists a neighbourhood around point Q, in which $L(v) < 0$. Denote by P the point whose co-ordinates are (x, y). Because of the continuity of function $f(P)$ one can find a semi-circular neighbourhood $\omega'_{x_0} \subset \omega_{x_0}$ around Q, for a given positive ε such that

$$f(Q) - \varepsilon \leqslant f(P) \leqslant f(Q) + \varepsilon. \tag{3.44}$$

throughout this neighbourhood.

Consider the two functions

$$\psi(P) = f(Q) + \varepsilon + kv(P), \tag{3.45}$$

$$\varphi(P) = f(Q) - \varepsilon - k_1 v(P), \tag{3.46}$$

where k and k_1, are as yet arbitrary positive numbers.

Starting from definite values of k and k_1 we have by (3.40), (3.45) and (3.46) that $L[\psi(P)] < 0$, $L[\varphi(P)] > 0$ everywhere in the neighbourhood ω'_{x_0}.

From (3.44) and (3.45) we find that in ω'_{x_0}, $\psi(P) \geqslant \varphi(P)$. In view of $v(P) > 0$ for $P \neq Q$ the number k in formula (3.45) can be chosen in such a way that on the semi-circle which is part of boundary ω'_{x_0}, we have $\psi(P) > M$.

The set of points in neighbourhood ω'_{x_0}, with ordinates $y > h$, constitute a domain which will be denoted by ω_h. On the boundary of ω_h, $\psi(P) \geqslant u_h(P)$. Therefore since everywhere in ω_h we have $L[\psi(P) - u_h(P)] < 0$, we can conclude that for every h and $P \in \omega_h$,

$$\psi(P) \geqslant u_K(P). \tag{3.47}$$

The number k_1 in (3.46) can be chosen in such a way that

$$\varphi(P) \leqslant u_h(P). \tag{3.48}$$

From (3.47) and (3.49) we conclude that

$$\varphi(P) \leqslant u(P) \leqslant \psi(P). \tag{3.49}$$

Passing to the limit $Q \to P$ we get from (3.29)

$$\varphi(Q) \leqslant u(Q) \leqslant \psi(Q)$$

or, noting (3.45) and (3.46)

$$f(Q) - \varepsilon \leqslant \lim_{P \to Q} u(P) \leqslant f(Q) + \varepsilon.$$

Hence it follows by the arbitrariness of ε, that $\lim\limits_{P \to Q} u(P) = f(Q)$.

In this way the function

$$u(x, y) = \lim_{i \to \infty} u_{h_i}(x, y)$$

is the unique solution of equation (3.36) which is regular in the domain D and coincides on its boundary $\Gamma = \sigma + AB$ with a given continuous function f.

In the case of equation (3.37), as proved by M. V. Keldish, the Dirichlet problem does not have always a solution for the above formulation of the problem.

The fact is that those solutions of equation (3.37) which do not have isolated singularities on $y = 0$, considered as a function of y for $y \to +0$, behave basically as if they were the solutions of the ordinary differential equation

$$y^m \varphi''(y) + b(x, 0)\varphi'(y) = 0.$$

This means that the solutions indicated can be characterized for $y \to +0$ as follows: (a) for $0 < m < 1$ all remain bounded; (b) for $m = 1$, $b(x, 0) < 1$, all remain bounded, but if $b(x, 0) \geqslant 1$ there exist unbounded solutions among them; (c) for $1 < m < 2$, $b(x, 0) \leqslant 0$ all remain bounded if $b(x, 0) > 0$, there are unbounded ones among them; (d) for $m \geqslant 2$, if $b(x, 0) < 0$, all remain bounded but if $b(x, 0) > 0$ there are unbounded ones among them. If $m \geqslant 2$, unbounded solutions exist also for $b(x, 0) \geqslant 0$. This can be explained by the fact that in this case the solutions of equation (3.37) in the neighbourhood of $y = 0$ behave as the solutions of the ordinary differential equation which, in addition to $y^m \varphi''(y) + b(x, 0)\varphi'(y)$, also contains terms of lower degree.

In all cases when the solutions of equation (3.37) are all bounded in the neighbourhood of $y = 0$ the Dirichlet problem as stated above has a solution which can be constructed in the same fashion as in the case of equation (3.36). To prove that the solution of equation (3.36) thus constructed assumes assigned continuous values on AB (on σ we do not need to check this fact) it is sufficient to prove that there is a barrier in all the cases considered which does not present any difficulties.

In fact we shall again look for the barrier in the form

$$v(x, y) = y^\beta + (x - x_0)^2, \quad 0 < x_0 < 1, \quad 0 < \beta < 1.$$

We have

$$L(v) = 2 + \beta(\beta - 1) y^{m+\beta-2} + 2a(x - x_0) + \beta b y^{\beta-1} + cv.$$

Hence it can be concluded that for sufficiently small values of y:

(1) for $0 < m < 1$ the sign of $L(v)$ is the same as the sign of $\beta(\beta - 1) y^{m+\beta-2}$, i. e. $L(v) < 0$; (2) for $m = 1$ and $b(x, 0) < 1$ the sign of $L(v)$ coincides with the sign of $\beta(\beta - 1 + b) y^{\beta-1}$ and if $\beta < 1 - b(x, 0)$, then $L(v) < 0$; (3) for $1 < m < 2$ and $b(x, 0) \leqslant 0$, if $\beta < 2 - m$, again $L(v) < 0$ and, finally, (4) for $m \geqslant 2$ and $b(x, 0) < 0$ the sign of $L(v)$ coincides with that of $\beta b y^{\beta-1}$, i.e. $L(v) < 0$.

Consequently, for a suitable choice of the neighbourhood ω_x, of point $Q(x_0, 0)$, $0 < x_0 < 1$ and of the power β, the expression $y^\beta + (x - x_0)^2$ can serve as barrier in all these cases.

If in the neighbourhood of $y = 0$ the solutions of equation (3.37) do not all remain bounded for $y \to 0$, the Dirichlet problem as stated above is incorrect.

It is found in these cases that the following problem (problem E in the terminology of M. V. Keldish) has a unique solution: find the solution $u(x, y)$ of equation (3.37) which is regular in domain D, remains bounded for $y \to 0$ and assumes given continuous values f only on the segment σ.

The solution for this problem can be constructed by the same methods as the solution of the Dirichlet problem.

Notice that in the case considered there is a positive function $w(x, y)$, which is regular in the closed domain $\bar{D}$, converges uniformly to infinity for $y \to 0$ and satisfies $L(w) < 0$ inside domain D.

For w we can take, for example, the expression

$$w(x, y) = -\ln y - (x - a)^n + k,$$

where the constant a is such that $x - a > 1$, the number n is a positive integer, k is a constant which is yet arbitrary and $\ln y$ signifies the main branch of this function.

Indeed, we have for $L(w)$ the expression

$$L(w) = y^{m-2} - n(n-1)(x-a)^{n-2} - \\ - y\, an\,(x-a)^{n-1} - \frac{b}{y} + cw.$$

For $m = 1$ and $b(x, 0) \geqslant 1$, the analyticity of $b(x, y)$, implies the existence of a positive number $A > 0$, such that

$$1 - b(x, y) < Ay.$$

Choose number n so that

$$n - 1 > 3|a|(x - a), \quad n(n-1) > 3A.$$

The number k is then chosen in such a way that function $w(x, y)$ becomes positive everywhere in D.

For these n and k

$$L(w) < A - \frac{1}{3} n[n - 1 + 3a(x-a)](x-a)^{n-2} - \\ - \frac{2}{3} n(n-1)(x-a)^{n-2} + cw < -\frac{1}{3} n(n-1) + cw < 0,$$

i.e. $w(x, y)$ satisfies all the required conditions.

If $1 < m < 2$, $b(x, 0) < 0$ or $m \geqslant 2$, $b(x, 0) \geqslant$ it is always possible to choose the number $A > 0$ in such a way that $y^{m-1} - b < Ay$. Therefore we can conclude by repeating the considerations quoted earlier that in these cases there exists a function $w(x, y)$ possessing the desired properties.

The uniqueness of the solution for problem E is now easily proved. For that it is sufficient to prove that the solution $u_0(x, y)$ for problem E, satisfying $u_0|_\sigma = 0$, vanishes identically in domain D.

Since on the one hand, given an arbitrary positive ε everywhere on the boundary Γ of domain D, $\varepsilon w \pm u_0 \geqslant 0$ and on the other hand inside D $L(w) < 0$ we can conclude that everywhere in D, $|u_0| \leqslant \varepsilon w$. From this by the arbitrariness of ε it follows that, everywhere in D, $u_0(x, y) = 0$.

The analyticity of the coefficients a, b and c was required for the sake of the simplicity of treatment. This requirement can, of course, be replaced by much weaker conditions concerning the smoothness of these functions.

It is to be noted that the Dirichlet problem for equation (3.37) was studied by Cibrario (3) in the special case when $m = 1$, $a = b = c = 0$.

§ 4. Some Other Problems and Generalizations

The results and the method of investigation established in the preceding paragraph remain in force for certain classes of elliptic differential equations in many-dimensional domains. For example the equations

$$z^m \left(\frac{\partial^2 u}{\partial x^2} + \frac{\partial^2 u}{\partial y^2} \right) + \frac{\partial^2 u}{\partial z^2} + A \frac{\partial u}{\partial x} + B \frac{\partial u}{\partial y} + C \frac{\partial u}{\partial z} + Du = 0$$

and

$$\frac{\partial^2 u}{\partial x^2} + \frac{\partial^2 u}{\partial y^2} + z^m \frac{\partial^2 u}{\partial z^2} + A \frac{\partial u}{\partial x} +$$

$$+ B \frac{\partial u}{\partial y} + C \frac{\partial u}{\partial z} + Du = 0, \quad D \leqslant 0,$$

where m is a positive number, are of this kind. Both of these equations are elliptic in the half-space $z > 0$ while the plane $z = 0$ is the curve of parabolic degeneracy.

The problem of finding solutions of equation (3.37) which are regular in domain D bounded in the closed domain $\bar{D}$ and subject to either the condition

$$\left.\frac{\partial u}{\partial \gamma}\right|_{\sigma} = f,$$

where γ is a given direction, or to the conditions

$$\left.\frac{\partial u}{\partial \gamma}\right|_{\sigma} = f, \quad u|_{AB} = \varphi,$$

were investigated in the papers of O. A. Oleinik (1) and N. D. Vedenskij (1).

The above mentioned problems concerning second order elliptic equations with parabolic degeneracy on the boundary of the domain were investigated by other methods also [cf. M. I. Vishick (2, 3), S. M. Michlin (1) L. D. Kudriavcev (1)].

There are a series of as yet uninvestigated questions concerning elliptic equations with parabolic degeneracy on the boundary of the considered domain.

1. Find the solution $u(x, y)$ of equation (3.37) which is regular in domain D but is not necessarily bounded for $y \to 0$ and satisfies the conditions

$$u|_{\sigma} = f, \quad \lim_{y \to 0} \varphi(x, y)\, u(x, y) = \varphi(x), \quad 0 \leqslant x \leqslant 1,$$

where f and φ are given continuous functions, and $\psi(x, y)$ is also a given function which becomes zero for $y \to 0$. (Some results concerning this problem can be found in the works of S. A. Tersanov (1), and Khoi Chun-i (1)].

2. Assume that on some parts of the segment AB contained in the boundary $\Gamma = \sigma + AB$ of domain D, the coefficient $b(x, 0)$ satisfies the following conditions: (a) $b(x, 0) < 1$ for $m = 1$; (b) $b(x, 0) \leqslant 0$ for $1 < m < 2$ (c) $b(x, 0) < 0$ for $m \geqslant 2$, and on the remaining parts of segment AB, conversely: (α) $b(x, 0) \geqslant 1$ for $m = 1$; (β) $b(x, 0) > 1$ for $1 < m < 2$; (γ) $b(x, 0) \geqslant 0$ for $m \geqslant 2$. Requiring that the solution $u(x, y)$

of equation (3.37) which is regular in domain D should remain bounded for $y \to 0$ (preserving, naturally, the uniqueness) clear up the question concerning the possibility of freeing those parts of AB, from the boundary conditions on which $b(x, 0)$ satisfies conditions (α), (β), (γ), regarding at the same time the values of $u(x, y)$ given on the remaining part of boundary Γ. For this problem cf. paper (1) of S. A. Tersenov where case $m = 1$ is considered.

3. Find the solution of equation (3.37) which is regular in D continuous in the closed domain $\bar{D}$ and satisfies the boundary conditions:

$$\left.\frac{\partial u}{\partial N}\right|_\sigma = f, \quad \left.\frac{\partial u}{\partial y}\right|_{AB} = \varphi, \tag{3.50}$$

or

$$u|_\sigma = f, \quad \left.\frac{\partial u}{\partial y}\right|_{AB} = \varepsilon, \tag{3.51}$$

where f and φ are given functions.

From the example of the equation

$$\frac{\partial^2 u}{\partial x^2} + y\frac{\partial^2 u}{\partial y^2} + \frac{1}{2}\frac{\partial u}{\partial y} = 0$$

it is clear that if the boundedness of $\partial u/\partial y$, for $y \to 0$, is demanded the problem of finding the function which is regular in domain D bounded in the closed domain $\bar{D}$ and satisfying only the condition $u|_\sigma = f$ has a unique solution.

Indeed by the change of variables $\xi = x$, $\eta = 2y^{1/2}$ the above equation becomes Laplace's equation

$$\frac{\partial^2 u}{\partial \xi^2} + \frac{\partial^2 u}{\partial \eta^2} = 0.$$

The requirement of boundedness on

$$\frac{\partial u}{\partial y} = \frac{2}{\eta}\frac{\partial u}{\partial \eta}.$$

when $y \to 0$, is equivalent to

$$\lim_{\eta \to 0}\frac{\partial u}{\partial \eta} = 0.$$

Therefore, the considered problem reduces to the Dirichlet problem for harmonic functions in domains symmetric with respect to $\eta = 0$.

In this way, the problem of this passage consists of the establishment of those m and $b(x, 0)$ for which problems (3.50) and (3.51) are possible and those m and $b(x, 0)$ for which segment AB has to be freed of boundary conditions, preserving the existence and uniqueness of the problem by means of fixing the boundary conditions on σ only.

A useful part may be played here by the following easily proved extremal principle: if $c(x, y) \leqslant 0$, $m \geqslant 1$, $b(x, 0) > 0$ the solution $u(x, y)$ of equation (3.37) which is regular in D continuous in the closed domain $\bar{D}$, and has bounded derivative $\lim\limits_{y \to 0} \partial u/\partial y$, cannot obtain its positive maximum or a negative minimum on the open segment AB.

4. Abandon the boundedness of $\partial u/\partial y$ for $y \to 0$ in stating problem 3 and replace the condition

$$\left.\frac{\partial u}{\partial y}\right|_{AB} = \varphi$$

by the much weaker condition

$$\lim_{y \to 0} \psi(x,y) \frac{\partial u}{\partial y} = \varphi,$$

where $\psi(x, y)$ is a given function vanishing for $y \to 0$.

5. The problem concerning the solution of elliptic systems in a domain on the boundary of which the type degenerates has not been investigated at all.

In the special case of the systems

$$y^m \frac{\partial^2 u}{\partial x^2} + \frac{\partial^2 u}{\partial y^2} + a \frac{\partial u}{\partial x} + b \frac{\partial u}{\partial y} + cu = 0$$

and

$$\frac{\partial^2 u}{\partial x^2} + y^m \frac{\partial^2 u}{\partial y^2} + a \frac{\partial u}{\partial x} + b \frac{\partial u}{\partial y} + cu = 0$$

$$(m > 0,\ a = \|a_{ik}\|,\ b = \|b_{ik}\|,\ c = \|c_{ik}\|),$$

which are elliptic for $y > 0$ and the type of which degenerates on the straight line $y = 0$ it would be interesting to study the Dirichlet problem in domain D under condition (3.34).

In these cases, there is, by (3.35) an integral representation analogous to (3.39) and therefore the method of the preceding paragraph can be used for constructing the desired solution but difficulties will arise when trying to elucidate the question whether the constructed solution assumes the given values for points on segment AB of boundary Γ of domain D.

CHAPTER 4

THE PROBLEM OF TRICOMI

THE linear partial differential equation of second order

$$y^{2m+1} \frac{\partial^2 u}{\partial x^2} + \frac{\partial^2 u}{\partial y^2} + a \frac{\partial u}{\partial x} + b \frac{\partial u}{\partial y} + cu = 0, \qquad (4.1)$$

where m is a non-negative integer, is elliptic for $y > 0$, hyperbolic for $y < 0$, but it has a parabolic degeneracy along the straight line $y = 0$.

In what follows a mixed domain will mean a domain which contains an interval of the axis $y = 0$.

The Cauchy problem in the hyperbolic part of the mixed domain with initial conditions on the line of parabolic degeneracy and the Dirichlet problem in the elliptic part of the domain were investigated for equation (4.1) in the preceding chapters.

It is natural to call the problem concerned with finding the solution of an equation of mixed type under one or another boundary conditions a mixed problem.

The first problem of this kind was stated and analysed by Tricomi (1—6) for the equation

$$y \frac{\partial^2 u}{\partial x^2} + \frac{\partial^2 u}{\partial y^2} = 0. \qquad (4.2)$$

Certain generalizations of the results of Tricomi concerning the equation

$$y^{2m+1} \frac{\partial^2 u}{\partial x^2} + \frac{\partial^2 u}{\partial y^2} = 0 \qquad (4.3)$$

were discovered by Gellerstedt (1), (2), (3).

M. A. Lavrent'ev suggested a simpler model for the equations of mixed type

$$\frac{\partial^2 u}{\partial x^2} + \operatorname{sgn} y \cdot \frac{\partial^2 u}{\partial y^2} = 0. \tag{4.4}$$

It will become clear from the later treatment that the equations of Tricomi and Lavrent'ev are fundamentally the same in the sense of the statement of the problem of mixed equations but the investigation of these problems in the case of the Lavrent'ev equation is significantly simpler.

§ 1. The Statement of the Problem of Tricomi

Let D be a singly connected finite mixed domain in the plane of the variables x, y, bounded by the simple Jordancurve σ with its endpoints at A (0, 0), B (1, 0), lying in the upper half-plane $y > 0$, and by the characteristics AC:

$$x - \frac{2}{3}(-y)^{3/2} = 0$$

and BC:

$$x + \frac{2}{3}(-y)^{3/2} = 1$$

of equation (4.2) (Fig. 5).

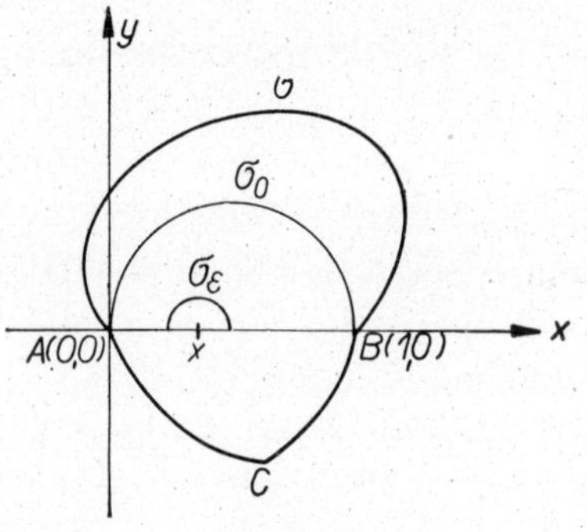

FIG. 5

The problem of Tricomi (problem T) consists of finding the function u (x, y) which is a solution of equation (4.2) in domain D; continuous in the closed domain $\bar{D}$ and assumes prescribed (continuous) values on σ and on AC:

$$u = \varphi \text{ or } \sigma, \tag{4.5}$$

$$u = \psi \text{ or } AC. \tag{4.6}$$

Denote again by D the mixed domain which is bounded by a simple Jordan curve σ lying in the upper plane $y > 0$ with its endpoints at $A\,(0, 0)$, $B\,(1, 0)$ and the characteristics AC: $x + y = 0$ and BC: $x - y = 1$ of equation (4.4) (Fig. 6).

The problem of Tricomi is stated for equation (4.4) in the following form: Find a function $u\,(x, y)$ which is continuous in the closed domain $\bar{D}$, has continuous derivatives $\partial u/\partial x$ and $\partial u/\partial y$, inside domain D satisfies equation (4.4) in domain D for $y \neq 0$ and assumes prescribed (continuous) values on the curve σ and on one of the characteristics; say on AC:

$$u = \psi \text{ on } AC. \tag{4.7}$$

(4.8)

According to the conditions for problem T, the function $\tau\,(x) = u\,(x, 0)$ must be continuous on segment $0 \leqslant x \leqslant 1$ for both equations (4.2) and (4.4). Apart from this, functions $\nu\,(x) = \partial u\,(x, y)/\partial y|_{y=0}$ and $\mathrm{d}\tau/\mathrm{d}x$ must be continuous and differentiable on the open segment $0 < x < 1$.

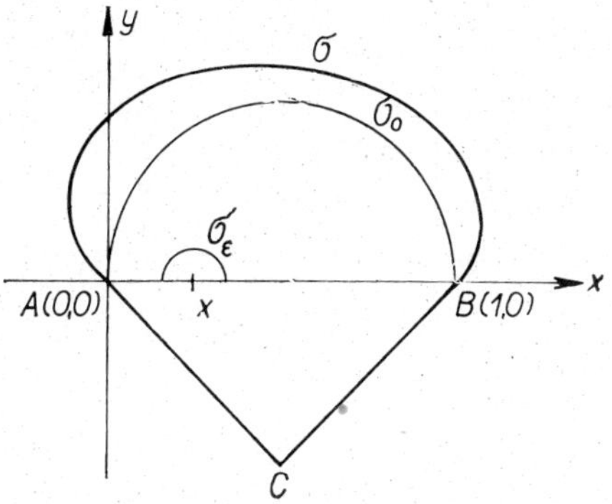

FIG. 6

At the same time it may happen that for $x \to 0$ and $x \to 1$ expression $\nu\,(x)$ becomes infinite not faster than of order $^2/_3$ (with respect to x and $1 - x$) in the case of equation (4.2). For equation (4.4) both $\nu\,(x)$, and $\mathrm{d}\tau/\mathrm{d}x$ may become infinite not faster than of order unity when $x \to 0$ and $x \to 1$.

The elliptic and hyperbolic sections of the mixed domain D will be designated in what follows by D_1 and D_2 respectively.

On the function ψ we impose the condition that along the whole characteristic AC it should possess continuous derivatives up to the third order.

In line with the conditions of problem T, function φ must be continuous. The restrictions on φ will be sharpened in §§ 3, 4 of this chapter.

§ 2. The Extremal Principle and the Uniqueness of the Solution of Problem T

Problem T will be investigated simultaneously for the Lavrent'ev and Tricomi equations. In both cases it can be assumed without restricting generality that $u(A) = u(B) = 0$.

The general solution of problem (4.4) which is continuous in the closed domain $\bar{D}_2$ and has continuous derivatives up to and including the second order inside D_2, is given by the familiar formula of D'Alembert

$$u(x, y) = f(x + y) + f_1(x - y), \tag{4.9}$$

where $f(t)$ and $f_1(t)$ are arbitrary continuous functions on $0 \leqslant t \leqslant 1$ twice continuously differentiable on $0 < t < 1$.

The general solution of equation (4.4) satisfying condition (4.8) can be promptly obtained for domain D_2, from formula (4.9):

$$u(x, y) = f(x + y) - f(0) + \psi\left(\frac{x - y}{2}\right).$$

Hence it follows directly that

$$\frac{\partial u}{\partial x} - \frac{\partial u}{\partial y} = \psi'\left(\frac{x}{2}\right), \quad y = 0, \quad 0 < x < 1 \tag{4.10}$$

or, what is the same thing, that

$$\sigma'(x) - \nu(x) = \psi'\left(\frac{x}{2}\right), \quad 0 < x < 1. \tag{4.11}$$

Identity (4.11) is the basic relation between $\tau(x)$ and $\nu(x)$ on segment AB, generated by the hyperbolic part D_2 of the mixed domain D.

The Darboux formula yielding the solution of the Cauchy problem for equation (4.2) in the hyperbolic part D_2 of domain D_2 with initial conditions $u(x, y)|_{AB} = \tau(x)$, $\partial u(x, y)/\partial y|_{AB} = \nu(x)$, has by § 3 of Chapter 2, the form

$$u(x, y) = \gamma_1 \int_0^1 \tau\left[x + \frac{2}{3}(-y)^{3/2}(2t-1)\right][t(1-t)]^{-5/6}\,dt +$$

$$+\left(\frac{4}{3}\right)^{2/3} \gamma_2\, y \int_0^1 \nu\left[x + \frac{2}{3}(-y)^{3/2}(2t-1)\right][(t(1-t)]^{-1/6}\,dt, \tag{4.12}$$

where

$$\gamma_1 = \frac{\Gamma(1/3)}{\Gamma^2(1/6)} = \frac{\sqrt[3]{4}\cdot\pi}{3\Gamma^3(1/3)},$$

$$\gamma_2 = \left(\frac{3}{4}\right)^{2/3} \frac{\Gamma(5/3)}{\Gamma^3(5/6)} = \frac{\sqrt[3]{6}\cdot\Gamma^3(1/3)}{4\pi^2}. \tag{4.13}$$

[formula (4.12) can be deduced in the same way as formula (2.47) with the difference that on this occasion $m = 1$ variable y is replaced by $-y$, and $\xi = x - 2/3\,(-y)^{3/2}$, $\eta = x + 2/3\,(-y)^{3/2}$].

Equating expression (4.12) on characteristics AC to function $\psi(x)$, we get

$$\gamma_1 \int_0^1 \tau(2tx)\,[t(1-t)]^{-5/6}\,dt - \gamma_2 (2x)^{2/3} \int_0^1 \nu(2tx)\,[t(1-t)]^{-1/6}\,dt,$$

$$0 \leqslant x \leqslant \frac{1}{2},$$

or

$$\gamma_1 x^{2/3} \int_0^x \frac{\tau(t)\,dt}{[t(x-t)]^{5/6}} - \gamma_2 \int_0^x \frac{\nu(t)\,dt}{[t(x-t)]^{1/6}} = \psi\left(\frac{x}{2}\right), \quad 0 \leqslant x \leqslant 1. \tag{4.14}$$

Applying the well known inversion formula

$$F(x) = \frac{\sin \pi a}{\pi} \frac{d}{dx} \int_0^x \frac{\Phi(t)\,dt}{(x-t)^{1-a}}$$

to the integral equation of Abel

$$\int_0^x \frac{F(t)\,\mathrm{d}t}{(x-t)^\alpha} = \Phi(x), \qquad 0 < \alpha < 1,$$

we get from formula (4.14) the expression

$$\tau(x) = \psi_1(x) + \gamma \int_0^x \frac{\nu(t)\,\mathrm{d}t}{(x-t)^{1/3}}, \tag{4.15}$$

for $\tau(x)$ where

$$\psi_1(x) = \frac{1}{2\pi\gamma_1} x^{5/6} \frac{\mathrm{d}}{\mathrm{d}x} \int_0^x \frac{\psi(t/2)}{t^{2/3}(x-t)^{1/6}}\,\mathrm{d}t,$$
$$\gamma = \frac{\gamma_2}{\gamma_1} \frac{\Gamma(5/6)}{\Gamma(1/6)\,\Gamma(2/3)} = \frac{3^{2/3}\Gamma^3(1/3)}{4\pi^2}. \tag{4.16}$$

Formula (4.15) yields for equation (4.2) the basic relation between $\tau(x)$ and $\nu(x)$, generated by the hyperbolic part D_2 of domain D.

Assuming that $\psi(x) = 0$, formulae (4.11) and (4.15) become

$$\tau'(x) - \nu(x) = 0, \tag{4.17}$$

$$\tau(x) = \gamma \int_0^x \frac{\nu(t)\,\mathrm{d}t}{(x-t)^{1/3}}. \tag{4.18}$$

Applying again the inversion formula of the Abel integral equation, (4.18) can be written in the form.

$$\nu(x) = \frac{\sqrt{3}}{2\pi\gamma} \frac{\mathrm{d}}{\mathrm{d}x} \int_0^x \frac{\tau(t)\,\mathrm{d}t}{(x-t)^{2/3}}. \tag{4.19}$$

It is easy to prove the following extremal principle for problem T: The solution $u(x, y)$ of problem T, which vanishes on characteristic AC, assumes both its positive maximum and negative minimum over the closed domain $\bar{D}_1$ on the curve σ for both equations (4.4) and (4.2).

Indeed, function $u(x, y)$ clearly cannot have an extremum inside domain D_1. Assume that the positive maximum for the closed domain $\bar{D}_1$ is assumed at an interior point $P(\xi, 0)$ of interval $0 < x < 1$. At point ξ, $0 < \xi < 1$ we clearly have the identity.

$$\tau'(\xi) = 0. \tag{4.20}$$

Consequently in the case of (4.4) we must have owing to (4.17),

$$\nu(\xi) = 0. \tag{4.21}$$

In the case of equation (4.2), owing to (4.20) equation (4.19) can be rewritten for ξ if $0 < x_0 < \xi$ in the form

$$\frac{2\pi\gamma}{\sqrt{3}} \nu(\xi) = \frac{2}{3} \int_0^{x_0} \frac{\tau(\xi) - \tau(t)}{(\xi - t)^{5/3}} dt +$$

$$+ \frac{2}{3} \int_{x_0}^{\xi} \frac{\tau(\xi) - \tau(t)}{(\xi - t)^{5/3}} dt + \tau(\xi) \frac{1}{\xi^{2/3}};$$

hence, since $\tau(\xi) > 0$, $\tau(\xi) - \tau(t) \geqslant$, we get

$$\nu(\xi) > 0. \tag{4.22}$$

Equation (4.21) and inequality (4.22) are contradictory. Lte $u(x, y)$ be a regular solution (non-trivial) in domain D^* of the equation

$$a \frac{\partial^2 u}{\partial x^2} + 2b \frac{\partial^2 u}{\partial x \partial y} + c \frac{\partial^2 u}{\partial y^2} +$$

$$+ a_1 \frac{\partial u}{\partial x} + b_1 \frac{\partial u}{\partial y} + c_1 u = 0, \qquad c_1 \leqslant 0,$$

with a positive definite form $dy^2 + 2\,b\,dx\,dy + c\,dy^2$ inside D^*. If at point P on the boundary Γ^* of domain D^* the function $u(x, y)$ assumes its extremal value, but contour Γ^* has the property that a circle k can be inscribed in domain D^*, through point P then at the said point the derivative in the direction of the centre of this circle with respect to the radius (if it exists) is different from zero, and in the case of maximum $\partial u/\partial r < 0$, and in the case of minimum $\partial u/\partial r > 0$.

This fact was noted for harmonic functions by Zaremba and for more general elliptic equations by Giraud [cf. Miranda (1), and also O. A. Oleinik (2)].

A simple proof of this fact is based on the existence of a Green's function for the Dirichlet problem in a circle. Therefore the proof is also valid for equation (4.1) a special case of which is the equation of Tricomi (4.2).

The extremal principle explained above is of great importance, first because the uniqueness of the solution of problem T, can be promptly deduced from it, secondly, because it makes it possible to construct an alternative procedure of the type of the well known Schwarz procedure for the solution of problem T under sufficiently general conditions.

The extremal principle in the form in which it is described here for problem T was first formulated and utilized in the case of the Lavrent'ev equation (4.4) in the papers of the present author (4—6). The same principle for the case of the Tricomi equation (4.2) was found a year later by Germain and Bader (1) [cf. also in the papers of the same authors (2,3)].

§ 3. Solution of Problem T by means of the Method of Integral Equations

We shall first establish the form of the functional relations between $\tau(x)$ and $\nu(x)$ on the segment $0 \leqslant x \leqslant 1$, which are generated by the elliptic part D_1 of the mixed domain for both the Lavrent'ev and the Tricomi equations.

Consider the fundamental solution $\ln |\zeta - z|$, $\zeta = \xi + i\eta$, $z = x + iy$, $\eta > 0$, $y > 0$, of equation (4.4). Denote by $G(x; \xi, \eta)$ the solution of equation (4.4) in the domain D_1 with a logarithmic singularity at point $\zeta = x$, $0 < x < 1$:

$$G(x; \xi, \eta) = -\ln|\zeta - x| + g(x; \xi, \eta),$$

where $g(x;\ \xi,\ \eta)$ is a regular harmonic function of ξ, η in the domain D_1 satisfying the conditions:

$$g(x; \xi, \eta) - \ln|\zeta - x| = 0, \quad \zeta \in \sigma,$$
$$\left.\frac{\partial g}{\partial \eta}\right|_{\eta=0} = 0. \tag{4.23}$$

Clearly $G(x;\ \xi,\ \eta)$ is the harmonic Green's function with logarithmic singularity at $\zeta = x$ for domain $D_1^* = D_1 + D^* + + AB$, where D_1^* is a domain which is the image in AB of the domain D_1.

Assume that curve σ satisfies the Liapunov condition; i.e. the tangent of the angle formed by the tangent of σ and a constant direction (for example the positive direction of the Ox axis) satisfies Hölder's condition; and that the partial derivatives $\partial u/\partial x$ and $\partial u/\partial y$ are continuous everywhere in domain D_1 apart from, perhaps at points A and B, where along certain directions, they might even become infinite in such a way as to preserve the validity of the integral formulae (*) and (**) below.

Corresponding restrictions must also be placed on the function φ. We shall return to them later in this paragraph.

Isolate point $(x, 0)$ from domain D_1 by the curve σ_ε of the semicircle $|\zeta - x| = \varepsilon$, $\eta \geqslant 0$ and apply to the remaining part of domain D_1 the formula of Green

$$\int \left(G\frac{\partial u}{\partial N} - u\frac{\partial G}{\partial N}\right) ds = 0, \tag{4.24}$$

where the length of curve s is measured from point B in the positive direction, and N is the inner normal of the boundary (cf. Fig. 6)

Taking the boundary condition (4.7) into account we shall have from (4.23) and (4.24)

$$\int_{\sigma_\varepsilon} \left(G\frac{\partial u}{\partial N} - u\frac{\partial G}{\partial N}\right) ds + \left(\int_0^{x-\varepsilon} + \int_{x+\varepsilon}\right) G\nu\, d\xi = \int_\sigma \varphi \frac{\partial G}{\partial N}\, ds.$$

Hence, in the limit, $\varepsilon \to 0$, we get

$$\tau(x) + \frac{1}{\pi}\int_0^1 [g(x;\xi,0) - \ln|\xi - x|]\,\nu(\xi)\,d\xi = \varphi_*(x), \tag{4.25}$$

where

$$\varphi_*(x) = \frac{1}{\pi}\int_\sigma \varphi \frac{\partial G}{\partial N}\,ds. \tag{4.26}$$

Formulae (4.11) and (4.25) provide the fundamental functional relations between $\tau(x)$ and $\nu(x)$ in the case of problem T for equation (4.4).

When σ coincides with the semi-circle $\sigma_0\colon |\zeta - 1/2| = 1/2$, $\eta \geqslant 0$, the function $G(x;\xi,\eta)$ can be expressed by the formula

$$G(x;\xi,\eta) = \ln\left|\frac{x + \zeta - 2x\zeta}{x - \zeta}\right|, \quad \zeta = \xi + i\eta$$

and relation (4.25) is significantly simpler:

$$\tau(x) - \frac{1}{\pi}\int_0^1 [\ln|t - x| - \ln(t + x - 2tx)]\,x(t)\,dt = \varphi_*(x). \tag{4.27}$$

By direct checking it is easy to see that function $|\zeta - x|^{1/}$ becomes a solution of equation (4.2) for $\zeta = \xi + 2/3\, i\eta^{3/2}$ $(\zeta \neq x,\ \eta < 0)$

Denote again by $G(x;\xi,\eta)$ the function of the form

$$G(x;\xi,\eta) = |\zeta - x|^{-1/3} + g(x;\xi,\eta),$$

where $g(x;\xi,\eta)$ is a solution of equation (4.2) regular with respect to ξ, and η in domain D_1, and satisfying the conditions:

$$\left.\begin{aligned} g &= -|\zeta - x|^{-1/3}, \quad \zeta \in \sigma, \\ \left.\frac{\partial g}{\partial \eta}\right|_{\eta=0} &= 0. \end{aligned}\right\} \tag{4.28}$$

The proof of the existence of function $G(x;\ \xi,\ \eta)$ does not cause any difficulty (the method of § 3, Chap. 3 needs some slight modification). In the special case when σ coincides with the so-called normal contour σ_0: $|\zeta - 1/2| = 1/2$, the function $G(x;\ \xi,\ \eta)$ is of the form:

$$G(x; \xi, \eta) = |\zeta - x|^{-1/3} - |\zeta + x - 2\zeta x|^{-1/3}.$$

Note that the normal contours $|\zeta - x| = \text{const}$, $\zeta = \xi + 2/3\, i\eta^{3/2}$ play the same part in the theory of Tricomi equations as the semi-circle $|\zeta - x| = \text{const}$, $\zeta = \xi + i\eta$ in the theory of the Lavrent'ev equation.

Isolating the point $(x, 0)$ by the curve σ_ε of the normal contour $|\zeta - x| = \varepsilon$, $\zeta = \xi + 2/3\ i\eta^{3/2}$ from domain D_1 (cf. Fig. 5) and applying Green's formula to the remaining part of D_1

$$\int \eta\left(\frac{\partial G}{\partial \xi} - G\frac{\partial u}{\partial \xi}\right) d\eta - \left(u\frac{\partial G}{\partial \eta} - G\frac{\partial u}{\partial \eta}\right) d\xi = 0,$$

we get in the limit $|\ \varepsilon \to 0$, by (4.28).

$$\lim\left(\int_0^{x-\varepsilon} + \int_{x+\varepsilon}^1\right) G\nu(\xi)\, d\xi - \int_\sigma u\left(\eta\frac{\partial G}{\partial \xi}\frac{\partial \xi}{\partial N} + \frac{\partial G}{\partial \eta}\frac{\partial \eta}{\partial N}\right) ds -$$

$$- \lim \int_{\sigma_\varepsilon} u\left(\eta\frac{\partial G}{\partial \xi}\frac{\partial \xi}{\partial N} + \frac{\partial G}{\partial \eta}\frac{\partial \eta}{\partial N}\right) ds = 0.$$

But

$$\lim_{\varepsilon \to 0} \int_{\sigma_\varepsilon} u\left(\eta\frac{\partial G}{\partial \xi}\frac{\partial \xi}{\partial N} + \frac{\partial G}{\partial \eta}\frac{\partial \eta}{\partial N}\right) ds = -\frac{1}{\gamma}\tau(x),$$

where

$$\gamma - 2^{-1/3}3^{-2/3}\int_0^\pi (\sin\vartheta)^{1/3}\, d\vartheta.$$

Therefore, noting boundary condition (4.5), we have finally

$$\tau(x) + \gamma\int_0^1 G(x; \xi, 0)\,\nu(\xi)\, d\xi = F^*(x), \tag{4.29}$$

where

$$F^* = \gamma \int_\sigma \varphi \left(\frac{\partial G}{\partial \xi} \frac{\partial \xi}{\partial N} + \frac{\partial G}{\partial \eta} \frac{\partial \eta}{\partial N} \right) \mathrm{d}s. \tag{4.30}$$

Formulae (4.15) and (4.29) are the fundamental relations between functions $\tau(x)$ and $\nu(x)$ for problem T in the case of equation (4.2).

In particular when σ_0: coincides with the normal contour $|\zeta - 1/2| = 1/2$, $\zeta = \xi + 2/3 i \eta^{3/2}$, $\eta \geqslant 0$, we get, instead of (4.29), the simpler relation

$$\tau(x) + \gamma \int_0^1 \left[\frac{1}{|t + x|^{1/3}} - \frac{1}{(t + x - 2tx)^{1/3}} \right] \nu(t)\, \mathrm{d}t = F^*(x). \tag{4.31}$$

It is clear that expressions (4.26) and (4.30) are analytic functions of x for $0 < x < 1$.

Eliminating $\tau(x)$ from (4.11) and (4.27) we get

$$\nu(x) - \frac{1}{\pi} \frac{\mathrm{d}}{\mathrm{d}x} \int_0^1 [\ln|t - x| - \ln(t + x - 2tx)]\, \nu(t)\, \mathrm{d}t = F(x), \tag{4.32}$$

where

$$F(x) = \varphi_*'(x) - 2 \frac{\mathrm{d}}{\mathrm{d}x} \psi\left(\frac{x}{2}\right). \tag{4.33}$$

Assume that x lies strictly inside the interval (0,1).

It is clear that

$$\frac{\mathrm{d}}{\mathrm{d}x} \int_0^1 \ln(t + x - 2tx)\, \nu(t)\, \mathrm{d}t = \int_0^1 \frac{1 - 2t}{t + x - 2tx} \nu(t)\, \mathrm{d}t. \tag{4.34}$$

On the other hand, for $\varepsilon \to 0$ we have

$$\lim I_\varepsilon(x) = \lim \int_0^{x-\varepsilon} \ln(x - t)\, \nu(t)\, \mathrm{d}t + \int_{x+\varepsilon}^1 \ln(t - x)\, \nu(t)\, \mathrm{d}t =$$

$$= I(x) = \int_0^1 \ln|t - x|\, \nu(t)\, \mathrm{d}t,$$

the limit existing uniformly in x.

Clearly the uniform limit

$$\lim_{\varepsilon \to 0} I'_\varepsilon(x) = \lim\,[\nu(x-\varepsilon) - \nu(x+\varepsilon)]\ln\varepsilon -$$

$$-\lim\left(\int_0^{x-\varepsilon} \frac{\nu(t)\,dt}{t-x} + \int_{x+\varepsilon}^{1} \frac{\nu(t)\,dt}{t-x}\right) = -\int_0^1 \frac{\nu(t)\,dt}{t-x},$$

exists, where the integral is (in the sense of Cauchy) the principal value. Therefore by a well known formula of classical analysis, we have

$$\frac{d}{dx}\int_0^1 \ln|t-x|\,\nu(t)\,dt = -\int_0^1 \frac{\nu(t)\,dt}{t-x}. \tag{4.35}$$

Because of (4.34) and (4.35) equation (4.32) assumes the following form:

$$\nu(x) + \frac{1}{\pi}\int_0^1 \left(\frac{1}{t-x} + \frac{1-2t}{t+x-2tx}\right)\nu(t)\,dt = F(x), \tag{4.36}$$

In this way the solution of problem T for the Lavrent'ev equation reduces to that of the singular integral equation (4.36) when σ is the semi-circle σ_0.

We return now to the functional relations (4.15) and (4.31). After eliminating $\tau(x)$ from them we get

$$\int_0^x \frac{\nu(t)\,dt}{(x-t)^{1/3}} + \int_0^1 \left[\frac{1}{|x-t|^{1/3}} - \frac{1}{(t+x-2tx)^{1/3}}\right]\nu(t)\,dt = \varphi_1^*(x), \tag{4.37}$$

where

$$\varphi_1^* = \frac{1}{\gamma}F^*(x) - \frac{1}{\gamma}\psi_1(x). \tag{4.38}$$

Applying the inversion formula of the Abel integral equation, we write equation (4.37) in the form

$$\nu(x) + \frac{\sqrt{3}}{2\pi}\,\frac{d}{dx}\int_0^1 \nu(t)\,dt\left[\int_0^x \frac{d\xi}{|\xi-t|^{1/3}(x-\xi)^{2/3}} -\right.$$

$$\left.-\int_0^x \frac{d\xi}{(\xi+t-2t\xi)^{1/3}(x-\xi)^{2/3}}\right] = \frac{3}{2}F(x), \tag{4.39}$$

where

$$F(x) = \frac{1}{\sqrt{3}\pi} \frac{d}{dx} \int_0^x \frac{\varphi_1^*(t)\, dt}{(x-t)^{2/3}} . \tag{4.40}$$

When x lies strictly inside the interval (0,1) the second integral term on the left side of formula (4.39) can be rewritten by the change of variables

$$\frac{x-\xi}{\xi - \dfrac{t}{2t-1}} = z^3, \qquad \frac{x-\xi}{\dfrac{t}{2t-1} - \xi} = z_1^3$$

in the form

$$\frac{-\sqrt{3}}{2\pi} \frac{d}{dx} \int_0^1 \nu(t)\, dt \int_0^x \frac{d\xi}{(\xi + t - 2t\xi)^{1/3} (x-\xi)^{2/3}} =$$

$$= -\frac{3\sqrt{3}}{2\pi} \frac{d}{dx} \left\{ \int_0^{1/2} \frac{\nu(t)\, dt}{(1-2t)^{1/3}} \int_0^{\left[\frac{x(1-2t)}{t}\right]^{1/3}} \frac{dz}{1+z^3} + \right. \tag{4.41}$$

$$\left. + \int_{1/2}^1 \frac{\nu(t)\, dt}{2t-1)^{1/3}} \int_0^{\left[\frac{x(2t-1)}{t}\right]^{1/3}} \frac{dz_1}{1-z_1^3} \right\} = -\frac{\sqrt{3}}{2\pi} \int_0^1 \left(\frac{t}{x}\right)^{2/3} \frac{\nu(t)\, dt}{t+x-2tx} .$$

Consider now the function

$$I_1(x) = \int_0^1 \nu(t)\, dt \int_0^x \frac{d\xi}{|\xi - t|^{1/3} (x-\xi)^{2/3}} = \lim I_{1\varepsilon}(x) =$$

$$= \lim_{\varepsilon \to 0} \left\{ \int_0^{x-\varepsilon} \nu(t)\, dt \left[\int_0^t \frac{(t-\xi)^{1/3} (x-\xi)^{2/3}}{d\xi} + \int_t^x \frac{d\xi}{(\xi - t)^{1/3} (x-\xi)^{2/3}} \right] + \right.$$

$$\left. + \int_{x+\varepsilon}^1 \nu(t)\, dt \int_0^x \frac{d\xi}{(t-\xi)^{1/3} (x-\xi)^{2/3}} \right\} .$$

By changing the variable of integration $\xi = (x-t)z + t$

it is easy to calculate the integral

$$\int_t^x \frac{d\xi}{(\xi - t)^{1/3}(x - \xi)^{2/3}} = \frac{2\pi}{\sqrt{3}}, \quad x > t.$$

On the other hand, changing the variable according to the formula $z^3 = (x - \xi)/(t - \xi)$ we get

$$\int_0^{\epsilon} \frac{d\xi}{(t - \xi)^{1/3}(x - t)^{2/9}} = -3 \int_{\left(\frac{x}{t}\right)^{1/3}}^{\infty} \frac{dz}{1 - z^3}, \quad t < x,$$

$$\int_0^x \frac{d\xi}{(t - \xi)^{1/3}(x - \xi)^{2/3}} = -3 \int_{\left(\frac{x}{t}\right)^{1/3}}^{0} \frac{dz}{1 - z^3} =$$

$$= -3 \int_{\left(\frac{x}{t}\right)^{1/3}}^{\infty} \frac{dz}{1 - z^3} + \frac{\pi}{\sqrt{3}}, \quad t > x.$$

Consequently

$$I_{1\varepsilon} = \frac{2\pi}{\sqrt{3}} \int_0^{x-\varepsilon} \nu(t)\, dt + \frac{\pi}{\sqrt{3}} \int_{x+\varepsilon}^{1} \nu(t)\, dt -$$

$$- 3\left(\int_0^{x-\varepsilon} + \int_{x-\varepsilon}^{1} \right) \nu(t)\, dt \int_{\left(\frac{x}{t}\right)^{1/3}}^{\infty} \frac{dz}{1 - z^3}.$$

Hence the existence of the uniform limit follows at once

$$I_1'(x) = \lim_{\varepsilon \to 0} I_{1\varepsilon}' = \frac{\pi}{\sqrt{3}} \nu(x(+ \int_0^1 \left(\frac{t}{x}\right)^{2/3} \frac{\nu(t)\, dt}{t - x}. \quad (4.42)$$

On the basis of (4.41) and (4.42) equation (4.39) can be written in the form

$$\nu(x) + \frac{1}{\pi\sqrt{3}} \int_0^1 \left(\frac{t}{x}\right)^{2/3} \left(\frac{1}{t - x} - \frac{1}{t + x + 2tx} \right) \nu(t)\, dt = F(x). \quad (4.43)$$

The singular integral equation (4.43) is equivalent to problem T for equation (4.2) in the case when σ coincides with the normal contour σ_0.

As was already shown the functions $\varphi_*(x)$ *a* and $F^*(x)$, defined by formulae (4.26) and (4.30) depend analytically on x for $0 < x < 1$. We now clear up the behaviour of these functions when $x \to 0$ and $x \to 1$.

Consider first the case when σ coincides with σ_0, and φ is continuous together with its first and second derivatives.

Note that function $a + bx + cy + dxy$ is a solution of equation (4.2) and function $a + bx + cy + dxy + e(x^2 - y^2 \operatorname{sgn} y)$ satisfies equation (4.4) for $y \neq 0$, and it is continuous together with its first derivatives on the straight line $y = 0$.

Noting this fact, we can assume without restricting the generality that

$$u(A) = u(B) = u'(A) = u'(B) = 0, \tag{4.44}$$

where the derivatives are along the direction of the tangent to the contour $\sigma_0 + AC$.

From (4.44) we see that the functions φ and ψ may be represented in the form

$$\varphi = \eta^2 \varphi_1(\zeta), \quad \psi = \eta^2 \psi_2(\zeta); \tag{4.45}$$

where ξ and η are the coordinates of the point ζ on contour σ_0.

For equation (4.4) we have

$$\left.\frac{\partial G}{\partial N}\right|_{\sigma_0} = \frac{2x(1-x)}{x^2 - (2x-1)\cos^2\dfrac{\vartheta}{2}}, \qquad 2\zeta - 1 = e^{i\vartheta}.$$

In accordance with this, we find by using (4.45) that the expression for $\varphi_*(x)$ assumes the form

$$\varphi_*(x) = \frac{x(1-x)}{\pi} \int_0^1 \varphi_1(t) \frac{t^{1/3}(1-t)^{1/2}}{x^3 - (2x-1)t}\, dt, \quad t = \cos^2\frac{\vartheta}{2}.$$

Noting (4.45) we get similarly from formulae (4.16) and (4.30)

$$\left.\begin{aligned} F^*(x) &= \frac{2\gamma}{\sqrt{9}}\, x(1-x) \int_0^1 \frac{\varphi_1(t)\, t^{1/3}(1-t)^{1/3}\, dt}{[x^2-(2x-1)\, t]^{7/6}}\,, \\ \psi_1(x) &= \frac{1}{2\pi\gamma_1}\left(\frac{3}{2}\right)^{4/3} x^{5/6}\, \frac{d}{dx}\int_0^x \frac{t^{2/3}\, \psi_2(t)\, dt}{(x-t)^{1/6}}\,. \end{aligned}\right\} \qquad (4.46a)$$

The function $\varphi_*(x)$ behaves in the same manner for $x \to 0$ and for $x \to 1$. Therefore we satisfy ourselves by examining its behaviour for $x \to 0$. For the derivative $\varphi_*'(x)$ we have from (4.46) the expression:

$$\varphi_*'(x) = \frac{1-2x}{\pi x^2}\, \varphi_1(t_1) \int_0^1 t^{1/2}(1-t)^{1/2}\left(1 - t\,\frac{2x-1}{x^2}\right)^{-1} dt -$$

$$- \frac{2(1-x)}{\pi x^2}\, \varphi_1(t_2) \int_0^1 t^{1/2}(1-t)^{1/2}\left(1 - t\,\frac{2x-1}{x^2}\right)^{-2} dt + \qquad (4.47)$$

$$+ \frac{2(1-x)}{\pi x^3}\, \varphi_1(t_3) \int_0^1 t^{3/2}(1-t)^{1/2}\left(1 - t\,\frac{2x-1}{x^2}\right)^{-2} dt.$$

The integrals on the right side can be expressed by the aid of the hypergeometric functions

$$\frac{\Gamma(b)\,\Gamma(c-b)}{\Gamma(c)}\, F\left(a, b, c, \frac{2x-1}{x^2}\right) =$$

$$= \int_0^1 t^{b-1}(1-t)^{c-b-1}\left(1 - t\,\frac{2x-1}{x^2}\right)^{-a} dt.$$

Noting the well-known formula

$$F\left(a, b, c, \frac{2x-1}{x^2}\right) =$$

$$= \left(\frac{x}{1-x}\right)^{2a} \frac{\Gamma(c)\,\Gamma(b-a)}{\Gamma(b)\,\Gamma(c-a)}\, F\left[a, c-b,\ 1+a-b, \left(\frac{x}{1-x}\right)^2\right] +$$

$$+ \left(\frac{x}{1-x}\right)^{2b} \frac{\Gamma(c)\,\Gamma(a-b)}{\Gamma(a)\,\Gamma(c-b)}\, F\left[b, c-a,\ 1-a+b, \left(\frac{x}{1-x}\right)^2\right],$$

we conclude that for $x \to 0$ the expression for $\varphi'_* (x)$ converges to a finite limit. By analogous considerations it is easy to see that $\varphi''_* (x)$ becomes infinite not faster than order unity. The same can be said about the behaviours of $\varphi'_* (x)$ and $\varphi''_* (x)$ for $x \to 1$.

Taking formulae (4.16) and (4.30) into account we conclude similarly that function $F^* (x) + \psi_1 (x)$ has a zero of the first order at $x = 0$, is continuous together with its first derivative in $0 \leqslant x \leqslant 1$, but its second derivative may become infinite at points $x = 0$ and $x = 1$ at a rate not greater than order $^2/_3$.

In this way we can draw the following conclusions under the above assumptions: (1) the right side of integral equation (4.36) is continuous in $0 \leqslant x \leqslant 1$, twice continuously differentiable in $0 < x < 1$, and its first derivative may become infinite at a speed not faster than order unity at points $x = 0$ and $x = 1$. (2) the right side of integral equation (4.43) is continuous in $0 \leqslant x \leqslant 1$ it has derivatives of first and second order in $0 < x < 1$, it has a zero of order $^1/_3$ at point $x = 0$, its first derivative may become infinite at point $x = 0$ not faster than of order $^2/_3$ and at $x = 1$ not faster than of order $^1/_3$.

We return now to the functional relations (4.11) and (4.25).

After eliminating $\tau (x)$ from these relations we get the integral equation

$$\nu (x) + \frac{1}{\pi} \int_0^1 \left(\frac{1}{t - x} + \frac{1 - 2t}{t + x - 2tx} \right) \nu (t) \, dt + \int_0^1 K (x, t) \, \nu (t) \, dt = F (x), \tag{4.48}$$

for the determination of function $\nu (x)$ where

$$K (x, t) = \frac{1}{\pi} \frac{\partial}{\partial x} [g (x; \; t. 0) - \ln (t + x - 2tx)]. \tag{4.49}$$

For $0 < x, t < 1$ the function $K (x, t)$ is continuously differentiable but at the endpoints of these intervals it may become infinite. In particular, however, whenever σ terminates in short portions AA' and BB' of the semi-circle σ_0, function $K (x, t)$

will have no singularities at the endpoints of the said intervals. In this case the same can be said about the behaviour of function $F(x)$ as about the right side of equation (4.36).

Eliminating the function $\tau(x)$ from (4.15) and (4.29) and repeatedly applying the transformations that were used for deducing integral equation (4.43) we get:

$$\nu(x) + \frac{1}{\pi\sqrt{3}} \int_0^1 \left(\frac{t}{x}\right)^{2/3} \left(\frac{1}{t-x} - \frac{2}{t+x-2tx}\right) \nu(t)\, dt + \\ + \int_0^1 K(x,t)\, \nu(t)\, dt = F(x), \tag{4.50}$$

where $K(x, t)$ is expressed through $g(x; t, 0)$.

With respect to the kernel $K(x, t)$ in equation (4.50) the same may be said as in the case of equation (4.48). Whenever σ terminates however in small sections AA' and BB' of the normal contour σ_0, function $K(x, t)$ is continuously differentiable for $0 \leqslant x \leqslant 1$, and the right sides of equations (4.43) and (4.50) behave identically.

If $\nu(x)$ can be determined from the integral equations obtained above then the values of $\tau(x)$ are given in $0 \leqslant x \leqslant 1$. Consequently the solution of the Dirichlet problem in domain D_1 and the solution of the Cauchy problem with given initial conditions

$$u(x, 0) = \tau(x), \qquad \left.\frac{\partial u(x,y)}{\partial y}\right|_{y=0} = \nu(x)$$

furnish a complete solution for problem T.

Remark.

The continuity of the third derivative of function $\psi(x)$ was demanded with a view to ensuring the existence in $0 < x < 1$ of the second derivative of function $F(x)$. However we shall later use only the fact that $F'(x)$ satisfies Hölder's condition and this always happens if $\psi''(x)$ satisfies Hölder's condition in $0 \leqslant x \leqslant 1$.

§ 4. Continuation. The Proof for the Existence of a Solution of the Integral Equations obtained in the Preceding Paragraph

We can easily find the solutions of the required form for integral equations (4.36) and (4.43), i.e. such solutions of these equations that are differentiable in $0 < x < 1$, permit a singularity of order not higher than unity in the case of equation (4.36) and of order not higher than $^2/_3$ in the case of equation (4.43) (cf. Tricomi (1)).

For this purpose let us consider the following auxiliary problem from the theory of holomorphic functions: determine the function $F(z) = u(x, y) + iv(x, y)$, which is holomorphic in the semi-circle $|2z - 1| < \operatorname{Im} z < 0$ continuous up to the boundary and satisfies the conditions:

$$\left.\begin{aligned} u\big|_{\sigma_0} = 0, \frac{\partial u}{\partial x} - \lambda \frac{\partial u}{\partial y} = f(x), \quad y = 0, \quad 0 < x < 1, \\ v(0, 0) = 0, \end{aligned}\right\} \tag{4.51}$$

where σ_0 is the semi-circle $|2z - 1| = 1$, $\operatorname{Im} z \geqslant 0$, λ is a real constant and $f(x)$ is a given function satisfying Hölder's condition. It is assumed that $\partial u/\partial x$ and $\partial u/\partial y$ are continuous inside the semi-circle up to the open interval $0 < x < 1$, $y = 1$, and may become infinite not faster than of order unity at points $z = 0$, $z = 1$.

The uniqueness of the solution for this problem is evident.

Acting in the same manner as when we deduced relation (4.27), and using the first of relations (4.51) we get

$$\begin{aligned} u(x, 0) - \frac{1}{\pi} \int_0^1 [\ln |t - x| - \\ - \ln (t + x - 2tx)] \left.\frac{\partial u(t, \eta)}{\partial \eta}\right|_{\eta=0} \mathrm{d}t = 0. \end{aligned} \tag{4.52}$$

After eliminating $\partial u(x, 0)/\partial x$ from (4.52) and from the second condition of (4.51) we have

$$\lambda \nu_1(x) + \frac{1}{\pi} \int_0^1 \left(\frac{1}{t - x} + \frac{1 - 2t}{t + x - 2tx} \right) \nu_1(t)\, \mathrm{d}t = -f(x), \tag{4.53}$$

where

$$\nu_1(x) = \frac{\partial u(x,y)}{\partial y}\bigg|_{y=0'} \qquad 0 < x < 1.$$

In this way the problem stated above can be reduced to the solution of integral equation (4.53) the solution $\nu_1(x)$ of which has to belong to the class of functions satisfying the Hölder condition in interval (0,1) and may perhaps possess singularities of order not higher than unity at the endpoints of this interval. Knowing the solution of equation (4.53) the solution of boundary problem (4.41) can be reduced in the familiar way to the Dirichlet problem. However, we shall treat it differently: we shall directly construct a solution for the problem indicated above and use it to obtain the solution of integral equation (4.53).

From the first of conditions (4.51) we conclude that the function $F(z)$ can be analytically continued across σ_0 to the whole upper half-plane

$$\begin{gathered}\frac{\partial u}{\partial x} + \lambda \frac{\partial u}{\partial y} = \frac{1}{(2x-1)^2} f\left(\frac{x}{2x-1}\right), \\ -\infty < x < 0, \quad 1 < x < \infty, \quad y = 0\end{gathered} \tag{4.54}$$

and, in addition, $F'(z)$ has a zero of second order at infinity.

Denote by $\Phi(z)$ the following function which is single-valued and holomorphic in the upper half-plane:

$$z^{\theta}(1-z)^{1-\theta} e^{iv} F'(z),$$

where $\vartheta = \tan^{-1}(-\lambda)$, $-\pi/2 < \vartheta < \pi/2$, $\Theta = -2\vartheta/\pi$ for $\lambda < 0$, $\Theta = (\pi - 2\vartheta)/\pi$ for $\lambda < 0$, and $z^{\theta}(1-z)^{1-\theta}$ is the branch of this function which is single-valued in the plane cut along $(-\infty, 0), (1, \infty)$ and is positive in $0 < z < 1$. Evidently $\Phi(\infty) = 0$.

By (4.54) and the second of conditions (4.51) we have

$$\operatorname{Re}\Phi(x) = \begin{cases} \dfrac{(-x)^{\theta}(1-x)^{1-\theta}}{\sqrt{(1+\lambda^2)}} f\left(\dfrac{x}{2x-1}\right) \dfrac{\operatorname{sgn}\lambda}{(2x-1)^2}, & -\infty < x < 0, \\ x^{\theta}(1-x)^{1-\theta} \dfrac{f(x)}{\sqrt{(1+\lambda^2)}}, & 0 < x < 1, \\ -\dfrac{x^{\theta}(x-1)^{1-\theta}}{\sqrt{(1-\lambda^2)}} f\left(\dfrac{\operatorname{sgn}\lambda}{2x-1}\right) \dfrac{x}{(2x-1)^2}, & 1 < x < \infty. \end{cases} \tag{4.55}$$

It is well-known [cf. for example N. I. Muschelishvili (1)], that the function $F_1(z) = u_1(x, y) + iv_1(x, y)$, which is holomorphic in the upper half-plane and converges to a finite limit for $z \to \infty$, can be expressed by means of the boundary values of its real part $u_1(t)$ on the real axis according to the Schwarz formula

$$F_1(z) = \frac{1}{\pi i} \int_{-\infty}^{\infty} \frac{u_1(t)\, dt}{t - z} + ib, \tag{4.56}$$

where $b = \text{Im } F_1(\infty)$, and it is assumed that function $u_1(t)$ satisfies certain conditions securing the existence of the improper integral on the right side of formula (4.56).

By Schwarz's formula (4.56) we can write

$$\Phi(y) = \frac{1}{\pi i} \int_{-\infty}^{\infty} \frac{\text{Re}\, \Phi(t)\, dt}{t - z}$$

or, noting (4.55)

$$F'(z) = \frac{e^{-i\vartheta}}{\pi i \sqrt{(1 + \lambda^2)}} \int_0^1 \left(\frac{t}{z}\right)^{\theta} \left(\frac{1 - t}{1 - z}\right)^{1-\theta} \left(\frac{1}{t - z} - \frac{\text{sgn}\, \lambda}{t + z - 2tz}\right) f(t)\, dt. \tag{4.57}$$

Hence after integrating we get

$$F(z) = \frac{e^{-i\vartheta}}{\pi i \sqrt{(1 + \lambda^2)}} \int_0^1 \left(\frac{z}{t}\right)^{1-\theta} \left(\frac{1 - z}{1 - t}\right)^{\theta} \left(\frac{1}{t - z} - \frac{\text{sgn}\, \lambda}{t + z - 2tz}\right) f_1(t)\, dt, \tag{4.58}$$

where

$$f_1 = \int_0^x f(t)\, dt.$$

Applying a well-known formula [c.f. N. I. Muschelishvili (1)], for the calculation of the limit values of the Cauchy-type integral for $z \to x$, $0 < x < 1$ and noting the identity

$$\frac{t}{t - x} + \frac{1 - 2t}{t + x - 2tx} = \frac{t}{x} \left(\frac{1}{t - x} - \frac{t}{t + x - 2tx}\right),$$

we get from formula (4.57) the solution (unique in the class of permitted solutions) of integral equation (4.53):

$$\nu_1(x) = -\frac{\lambda}{1+\lambda^2} f(x) + \frac{1}{\pi(1+\lambda^2)} \int_0^1 \left[\frac{x(1-t)}{t(1-x)}\right]^{1-\theta} \left(\frac{1}{t-x} + \frac{1-2t}{i+x-2tx}\right) f(t)\,dt \tag{4.59}$$

for $\lambda > 0$ and

$$\nu_1(x) = -\frac{\lambda}{1+\lambda^2} f(x) + \frac{1}{\pi(1+\lambda^2)} \int_0^1 \left[\frac{t(1-x)}{x(1-t)}\right]^{\theta} \left(\frac{1}{t-x} + \frac{1-2t}{t+x-2tx}\right) f(t)\,dt$$

for $\lambda < 0$.

In particular when $\lambda = 1$, $f(x) = -F(x)$ and $\nu(x) = \nu(x)$, equation (4.53) coincides with equation (4.36), the inversion formula of which is yielded by (4.59) [cf. A. V. Bitsadze (4—7)]:

$$\nu(x) = \frac{1}{2} F(x) - \frac{1}{2\pi} \int_0^1 \sqrt{\left(\frac{x(1-t)}{t(1-x)}\right)} \left(\frac{1}{t-x} + \frac{1-2t}{t+x-2tx}\right) F(t)\,dt. \tag{4.60}$$

If $\lambda = \sqrt{3}$, $\nu = x^{-1/3}\nu(x)$, $f(x) = -x^{-1/3} F(x)/\sqrt{3}$, then equation (4.53) coincides with equation (4.43), the solution of which is again furnished by formula (4.59) [cf. Tricomi (1) and S. G. Michlin (2)]:

$$\nu(x) = \frac{3}{4}\left\{F(x) - \frac{1}{\pi\sqrt{3}} \int_0^1 \left[\frac{t(t-t)}{x(1-x)}\right]^{1/3} \left(\frac{1}{t-x} - \frac{1}{(t+x-2tx)}\right) F(t)\,dt\right\}. \tag{4.61}$$

Applying inversion formulae (4.60) and (4.61) the integral equations (4.48) and (4.50) can be rewritten in the form

$$\nu(x) + \int_0^1 K_1(x,t)\,\nu(t)\,dt = F_1(x). \tag{4.62}$$

Whenever σ terminates arbitrarily small portions AA' and BB' of curve σ_0, the kernel $K_1(x, t)$ of integral equation (4.62) may have only non-removable singularities of integrable order at $x = 0$, $t = 0$, $x = 1$, $t = 1$. Therefore in this case problem T, for both equations (4.4) and (4.2) can be reduced to an equivalent Fredholm integral equation of the second kind, the solvability of which follows from the uniqueness of this problem.

If curve σ does not satisfy the conditions described above kernal $K(x, t)$ in equations (4.48) and (4.50) may itself be singular at $x = 0$, $t = 0$, $x = 1$, $t = 1$, in general depending on the angles between σ and axis Ox at the points A and B. Therefore the integral equation (4.62) must be further investigated in that case.

The function $\nu(x)$, which is a solution of integral equations (4.36) and (4.61)) actually satisfies all the restrictions put on $\partial u(x, y)/\partial y_{y=0}$ in § 1 of the present chapter. In particular, the differentiability of $\nu(x)$ in $0 < x < 1$ can be checked by integrating by parts in formula (4.60) and (4.61) the properties of function $F(x)$, which were established in the preceding paragraph make this procedure perfectly legitimate. From the same formula it follows that $v(x)$ has no singularity at $x = 0$ but it can actually become infinite of order $^1/_2$ and $^1/_3$ respectively at $x = 1$.

§ 5. Other Methods for Solving Problem T

In this paragraph we shall be speaking mainly about problem T for equation (4.4).

Let us first assume that curve σ satisfies the Liapunov condition, the partial derivatives $\partial u/\partial x$ and $\partial u/\partial y$ of the desired solution $u(x, y)$ of problem T are continuous in domain D right up to the open curve σ but they may become infinite, not faster than of order unity, at points A and B.

With these assumptions we get from (4.10) after integrating

$$u(x, 0) + v(x, 0) = 2\psi\left(\frac{x}{2}\right), \quad 0 \leqslant x \leqslant 1, \qquad (4.63)$$

where $v(x, y)$ is a function harmonic in domain D_1 and conjugate to function $u(x, y)$.

If we can find solutions $u_1(x, y)$ and $u_2(x, y)$ for problem T, satisfying the respective conditions

$$u_1|_\sigma = \varphi, \quad u_1|_{AC} = 0, \tag{4.64}$$

$$u^2|_\sigma = 0, \quad u_2|_{AC} = \psi, \tag{4.65}$$

the desired solution of problem T, satisfying conditions (4.7) and (4.8) can be represented in the form of the sum

$$u(x, y) = u_1(x, y) + u_2(x, y).$$

Denote by $F_1(z)$ the function $u_1(x, y) + iv_1(x, y)$, which is holomorphic in domain D_1 and satisfies condition $F_1(0) = 0$. From (4.63) and the second of conditions (4.64), we have $\operatorname{Re}(1 - i) F_1(x) = 0$, $0 \leqslant x \leqslant 1$. Therefore the function $F_1(z)$ can be analytically continued from domain D_1 across the segment AB into the domain D_1^*, (which is the image of D_1 in AB) so that

$$F_1(z) = \begin{cases} u_1(x, y) + iv_1(x, y), & y \geqslant 0, \\ -v_1(x, -y) - iu_1(x, -y), & y \leqslant 0. \end{cases} \tag{4.66}$$

We denote by D^* the union of domains D_1 and D_1^* together with the open interval AB.

From the first of conditions (4.64) and (4.66) we get the boundary conditions:

$$\operatorname{Re} F_1(t)|_\sigma = \varphi(t), \quad \lim F_1(t)|_{\bar{\sigma}} = -\varphi(t), \tag{4.67}$$

for function $F_1(z)$ where $\bar{\sigma}$ is the mirror image of σ in the real axis.

In this way to find a function $u(x, y)$ which is harmonic in the domain D_1 it is sufficient to find a function $F_1(z)$, which is holomorphic in domain D^*, continuous in the closed domain $\bar{D}^*$ and satisfies the boundary conditions (4.67), and the further condition

$$F_1(0) = 0. \tag{4.68}$$

Since D^* is symmetric with respect to the real axis and contains the segment AB, it can be mapped conformally onto the circle $|z - 1/2| < 1/2$ in such a way that σ becomes a one-to-one continuous image of the upper semi-circle σ_0, and $\bar{\sigma} - c$ of the lower semi-circle of this circle. For this reason it can be assumed from the beginning that σ and the semi-circle σ_0 coincide.

With a view to finding a function $F_1(z)$, which is holomorphic in circle $|z - 1/2| < 1/2$ and satisfies conditions (4.67) and (4.68), we introduce the holomorphic function:

$$\Phi_1(z) = \mathrm{e}^{-i\pi/4} \sqrt{\left(\frac{z}{1-z}\right)} F_1(z),$$

where the radical indicates that branch of this function for which $\sqrt{\left(\frac{z}{1-z}\right)} = i + \ldots$ (for large $|z|$).

For the function $\Phi(z)$ we have the boundary conditions

$$\left.\begin{aligned} \operatorname{Re} \Phi_1 |_{\sigma_0} &= \mathrm{e}^{-i\pi/4} \sqrt{\left(\frac{t}{t-t}\right)} \varphi(t), \\ \operatorname{Re} \Phi_1 |_{\bar{\sigma}_0} &= \mathrm{e}^{-i\pi/4} \sqrt{\left(\frac{1 - \frac{t}{2t-1}}{\frac{t}{2t-1}}\right)} \varphi\left(\frac{2t-1}{t}\right). \end{aligned}\right\} \quad (4.69)$$

However the function $\Phi_1(z)$, which is holomorphic in the circle $|z - 1/2| < 1/2$ and is such that $\Phi_1(0) = 0$, is determined uniquely by means of the boundary conditions for its real part according to the formula of Schwarz:

$$\Phi_1(z) = \frac{1}{\pi i} \int_{\sigma_0 + \bar{\sigma}_0} \frac{z}{t} \frac{\operatorname{Re} \Phi(t)\, \mathrm{d}t}{t - z}. \quad (4.70)$$

Taking (4.69) into account we get from formula (4.70), after a simple change of the variable of integration

$$F_1(z) = \frac{1}{\pi i} \int_{\sigma_0} \sqrt{\left(\frac{z(1-z)}{t(1-t)}\right)} \left(\frac{1}{t-z} - \frac{1}{t+z-2tz}\right) \varphi(t)\, \mathrm{d}t.$$

The real part of function $F_1(z)$ furnishes the required function $u_1(x, y)$ in domain D_1. We also have in domain D_2 the expression:

$$u_1(x, y) = u_2(x + y, +0) \tag{4.71}$$

for function $u_1(x, y)$. It is clear from formula (4.71) that for $0 < x + y < 1$ function $u_1(x, y)$ is infinitely often differentiable.

Repeating the considerations just made, it is easy to see that the solution $u_1(x, y)$ of problem T with zero initial conditions on characteristic AC has in the semi-circle k: $|t - \xi| < \varepsilon$, $\operatorname{Im} t \geqslant 0$, $0 < \xi < 1$, the representation

$$u_1(x, y) = \operatorname{Re}\left\{\frac{1}{\pi i}\int\limits_{C_k}\frac{\sqrt{z - \xi + \varepsilon)(\xi + \varepsilon - z)}}{\sqrt{(t - \xi + \varepsilon)(\xi + \varepsilon - t)}}\frac{u(t)\,dt}{t - z} - \right.$$
$$\left. - \frac{1}{\pi}\int\limits_{\overline{C}_k}\frac{\sqrt{(z - \xi + \varepsilon)(\xi + \varepsilon - z)}}{\sqrt{(t - \xi + \varepsilon)(\xi + \varepsilon - t)}}\frac{u(\bar{t})\,dt}{t - z}\right\}$$

where C_k and $\overline{C}_k$ are the upper and lower semi-circles of circle k.

This formula implies a peculiar mean value theorem

$$u_1(\xi, 0) = \frac{1}{\pi}\int\limits_0^{\pi}\sqrt{\left(\frac{1}{2}\tan\frac{\vartheta}{2}\right)}u(\vartheta)\,d\vartheta, \quad t - \xi = \varepsilon e^{i\vartheta},$$

whence, in turn, the extremal principle, proved in § 2 follows at once.

Note, that without restricting the generality it can be assumed that $u|_\sigma = 0$. Indeed, denote by $u_0(x, y)$ the function which is harmonic in domain D_1 and satisfies the conditions:

$$u_0|_\sigma = \varphi, \quad \left.\frac{\partial u_0}{\partial y}\right|_{y=0} = 0, \quad 0 < x < 1. \tag{4.72}$$

Due to the second of conditions (4.72) function $u_0(x, y)$ can be continued harmonically across AB into domain D_1^*, so that for $z = x + iy \in D_1^*$, $u_0(x, y) = u_0(x - y)$. Therefore $u_0(x, y)$ can be obtained as the solution of the Dirichlet problem for

D^*. We are interested in the values of the harmonic function $u_0(x, y)$ only in domain D_1. Let $u_0(x, +0) = \tau_0(x)$. Evidently function $w(x, y)$, which is equal to $u_0(x, y)$ in domain D_1, and to $1/2\,[\tau_0(x + y) + \tau_0(x - y)]$ in domain D_1, is the solution of problem D_2 satisfying the conditions:

$$w|_\sigma = \varphi, \quad w|_{AC} = \frac{1}{2}[\tau_0(0) + \tau_0(2x)].$$

In this way if $u(x, y)$ is the solution of problem T, satisfying conditions (4.7) and (4.8) then the difference $u(x, y) - w(x, y) = u_2(x, y)$ will be a solution of problem T, satisfying conditions:

$$u_2|_\sigma = 0 \quad u_2|_{AC} = \psi(x) - \frac{1}{2}[\tau_0(0) + \tau_0(2x)] \equiv \psi_1(x).$$

Thus, we have due to (4.10):

$$\frac{\partial u_2}{\partial x} - \frac{\partial u_2}{\partial y} = 2\frac{\mathrm{d}}{\mathrm{d}x}\psi\left(\frac{x}{2}\right), \quad y = 0, \quad 0 < x < 1.$$

Denote by $F_2(z)$ the function $u_2(x, y) + iv_2(x, y)$ which is holomorphic in domain D_1 and satisfies the conditions:

$$\left.\begin{aligned} \operatorname{Re} F_2(z)|_\sigma = 0, \quad \operatorname{Re}(1 - i)F_2'(z) = 2\frac{\mathrm{d}}{\mathrm{d}x}\psi\left(\frac{x}{2}\right), \\ \operatorname{Im} F_2(0) = 0. \end{aligned}\right\} \tag{4.73}$$

In this way the problem of finding by means of conformal mapping function $F_2(z)$, which is holomorphic in domain D_1 and satisfies conditions (4.73), reduces to the case when σ coincides with semi-circle σ_0. But in this case the expression for $F_2(z)$ can be obtained directly from formula (4.58) by putting $\lambda = 1$ and $f_1 = 2\psi(t/2)$:

$$\begin{aligned} F_2(z) = \frac{2}{\pi(1 + i)}\int_0^1 \sqrt{\frac{z(1 - z)}{t(1 - t)}}\left(\frac{1}{t - z} - \right. \\ \left. - \frac{1}{t + z - 2tz}\right)\psi\left(\frac{t}{2}\right)\mathrm{d}t. \end{aligned} \tag{4.74}$$

By separating the real part of (4.74) we get the values of the required solution $u_2(x, y)$ of problem T for domain D_1. The values of $u_2(x, y)$ in domain D_2 are furnished by the formula:

$$u_2(x, y) = u_2(x + y, +0) - \psi\left(\frac{x+y}{2}\right) + \psi\left(\frac{x-y}{2}\right). \quad (4.75)$$

It is clear from (4.74) and (4.75) that the degree of smoothness possessed by $u_2(x, y)$ in domain D_2 depends on the degree of smoothness possessed by function $\Psi(x)$.

The solution of problem T for equation (4.4) under the condition that φ is a differentiable function can be reduced directly to the Dirichlet problem (M. A. Lavrent'ev and A. B. Bitsadze (1)].

Indeed for discovering the required solution $u(x, y)$ of problem T in domain D it is sufficient to find a function $u(x, y)$, which is harmonic in domain D_1 and satisfies boundary conditions (4.5) and (4.10). Use the conformal mapping $z = f(z_1)$, $z_1 = x_1 + iy_1$ of domain D_1 into sector $0 < \arg z_1 < \pi/4$, which maps curve σ into radius $\arg z_1 = 0$, segment AB into radius $\arg z_1 = \pi/4$ and points A and B into ∞ and 0. The derivative of $u(x, y)$ in the direction of the tangent of σ goes over into the derivative along the negative axis Ox_1. Therefore, for the harmonic function $u[f(z_1)] = u_1(x_1, y_1)$, which is the image with this transformation of function $u(x, y)$, the values of derivative $\partial u_1/\partial x_1$ will be known due to (4.5) and (4.10) on the boundary of section $0 < \arg z_1 < \pi/4$, and this is precisely the Dirichlet problem for the harmonic function $\partial u_1/\partial x_1$.

On the basis of the extremal principle proved in §2 the existence of the solution for problem T can be asserted without the restrictions which were imposed in the preceding paragraph, and at the beginning of the present paragraph. Namely, it will be proved that problem T has a solution if σ is a smooth Jordan curve and the function φ which is given on it is continuous.

Denote by σ_1 the smooth Jordan curve with endpoints in A and B, which satisfies the Liapunov condition and lies entirely inside in the elliptic part D_1 of domain D. Let Δ be the mixed domain bounded by curve σ_1 and characteristics AC and BC of equation (4.4). We shall assume (and this is not a restriction on generality) that σ_0 and σ have common tangents near points A and B.

Construct two sequences of functions

$$u_1(x, y), \ldots, u_n(x, y), \ldots \tag{4.76}$$

and

$$v_1(x, y), \ldots, v_n(x, y), \ldots \tag{4.77}$$

with the following properties: $u_1(x, y)$ is a solution of problem T in domain Δ:

$$u_1|_{\sigma_1} = 0 \quad u_1|_{AC} = \psi(x). \tag{4.78$_1$}$$

Function $v_1(x, y)$ is harmonic in D_1 and satisfies the conditions:

$$v_1|_\sigma = \varphi, \; v_1 = u_1 \text{ on } AB. \tag{4.79$_1$}$$

Functions $u_n(x, y)$, $n = 2, 3, \ldots$, are solutions of problem T in domain Δ:

$$u_n = v_{n-1} \text{ on } \sigma_1, \; u_n = \psi(x) \text{ on } AC, \tag{4.78$_2$}$$

and $v_n(x, y)$, $n = 2, 3, \ldots$ are harmonic functions in D_1 satisfying the conditions

$$v_n|_\sigma = \varphi, \; v_n = u_n \text{ on } AB. \tag{4.79$_2$}$$

Apart from these, assume that $\partial u_n/\partial x$ and $\partial u_n/\partial y$, $n = 1, 2., \ldots$, are everywhere continuous in the closed domain $\bar{\Delta}$ apart from perhaps at points A and B (near to which, as it was already mentioned in the statement of problem T, $\partial u_n/\partial x$ and $\partial u_n/\partial y$ may have singularities of order not higher than unity).

Functions u_n and v_n can be constructed.

Denote by N the larger of $\max\limits_{(x,y)\in\Delta} |u_1(x, y)|$ and $\max\limits_{(x,y)\in D_1} |v_1(x, y)|$.

We conclude from conditions (4.78$_1$) and (4.78$_2$) that

$$|u_2(x, y) - u_1(x, y)| = |v_1(x, y)| \leqslant N \text{ on } \sigma_1. \tag{4.80}$$

However, on the other hand we have by (4.78$_1$) and (4.78$_2$) that

$$u_2 - u_1 = 0 \text{ on } AC.$$

Therefore the extremal principle can be applied to the difference $u_2(x, y) - u_1(x, y)$ due to which we get according to (4.80) the estimate

$$|u_2(x, y) - u_1(u_1, y)| \leqslant N, \; (x, y) \in \Delta.$$

From (4.79_1) and (4.79_2) we have

$$v_2 - v_1 = 0 \text{ on } \sigma, \; v_2 - v_1 = u_2 - u_1 \text{ on } AB.$$

Hence, according to the familiar properties of harmonic functions we have

$$|v_2 - v_1| \leqslant Nq, \; 0 < q < 1, \text{ on } \sigma_1.$$

Repeating this consideration again and again, we get

$$|u_n - u_{n-1}| \leqslant Nq^{n-2} \text{ on } \sigma_1 \text{ and on } AB, \tag{4.81}$$

$$|v_n - v_{n-1}| \leqslant Nq^{n-2} \text{ on } AB, \; |v_n - v_{n-1}| \leqslant Nq^{n-1} \text{ on } \sigma_1. \tag{4.82}$$

On the basis of (4.81) and (4.82) one can conclude that series

$$u_1 + (u_2 - u_1) + (u_3 - u_2) + \dots, \tag{4.83}$$

$$v_1 + (v_2 - v_1) + (v_3 - v_2) + \dots \tag{4.84}$$

converge absolutely and uniformly to harmonic functions $u(x, y)$ and $v(x, y)$ respectively in the elliptic parts of domain Δ and D.

Since $u_n = v_{n-1}$ on σ_1 and $u_n = v_n$ on AB, hence follows that $\lim_{n\to\infty} u_n(x, y) = \lim_{n\to\infty} v_n(x, y) = u(x, y) = v(x, y)$ in the elliptic part of domain Δ. Therefore, $v(x, y)$ is the analytic continuation of $u(x, y)$ from the elliptic part of domain Δ into the elliptic part of domain D and $v(x, y) = \varphi$ on σ.

In view of the fact that $u_n(x, y) - u_{n-1}(x, y) = 0$ on AC, we have in the hyperbolic part of Δ the formula

$$u_n(x, y) - u_{n-1}(x, y) = u_n(x + y, +0) - u_{n-1}(x + y, +0).$$

Hence, we have also the absolute and uniform convergence of series (4.83) in the hyperbolic part of Δ. Since $u_n(x, y) = \psi(x)$ on AC, $u(x, y) = \lim u_n(x, y) = \psi(x)$ on AC.

Consider now function $u_n^*(x, y)$, which is the harmonic conjugate of function $u_n(x, y)$ in the elliptic part of domain Δ, and such that $u_n^*(0, 0) = 0$.

From equations $u_n(x, y) - u_{n-1}(x, y) = 0$ valid on AC follows $u_n(x, 0) - u_{n-1}(x, 0) + u_n^*(x, 0) - u_{n-1}^*(x, 0) = 0, 0 < x < 1$.

It follows in turn, that the analytic continuation of $u_n(x, y) - u_{n-1}(x, y)$ into the domain which is the symmetric image of the elliptic part of domain Δ with resepect to the real axis is the function $u_n^*(x, -y) \pm u_{n-1}^*(x, y)$. We thus conclude that the domain in which the series $u_2 - u_1 + (u_3 - u_2) + \ldots$ converges can be extended outside the elliptic part of domain Δ below AB. Therefore the derivatives

$$\left(\frac{\partial u}{\partial y}\right)_{y=+0} = \nu(x) = \lim_{n\to\infty} \nu_n(x) = \lim_{n\to\infty} \left(\frac{\partial u_n}{\partial y}\right)_{y=+0}$$

and

$$\left(\frac{\partial u}{\partial x}\right)_{y=+0} = \tau'(x) = \lim_{n\to\infty} \tau_n'(x) = \lim_{n\to\infty} \left(\frac{\partial u_n}{\partial x}\right)_{y=+0}$$

exist.

But in the hyperbolic part of domain Δ we have

$$u_n(x, y) = \frac{1}{2}\tau_n(x+y) + \frac{1}{2}\tau_n(x-y) - \frac{1}{2}\int_{x+y}^{x-y} \nu_n(t)\,\mathrm{d}t, \tag{4.85}$$

where

$$\left(\frac{\partial u_n}{\partial y}\right)_{y=+0} = \nu_n(x), \qquad \left(\frac{\partial u_n}{\partial x}\right)_{y+=0} = \tau_n'(x). \tag{4.86}$$

By passing to the limit we get from (4.85) and (4.86)

$$u(x, y) = \frac{1}{2}\tau(x+y) + \frac{1}{2}\tau(x-y) - \frac{1}{2}\int_{x+y}^{x-y} \nu(t)\,\mathrm{d}t \tag{4.87}$$

and

$$\left(\frac{\partial u}{\partial x}\right)_{y=-0} = \tau'(x), \quad \left(\frac{\partial u}{\partial y}\right)_{y=-0} = \nu(x).$$

In this way function $u(x, y)$ which is the sum of series (4.83) in the elliptic part of domain D, and is represented by formula (4.87) in the hyperbolic part of this domain is the solution of problem T.

The above quoted alternative method for solving problem T is contained in the thesis of the author (5) and (6) which was defended in the spring of 1951. This method is applicable without change for solving problem T in the case of the Tricomi equation (4.2) (cf. Germain and Bader).

The working out of constructive methods for finding the approximate solutions of problem T is of great importance.

The approximate solution of problem T can be found by the method of finite differences for both equations (4.2) and (4.4). [cf. for example the papers of Z. I. Halilov, (1) V. G. Karmaov (1), (2), O. A. Ladizhenskii, and A. G. Filippov (1)] but we shall spend no more time on this here.

§ 6. Examples and Generalizations

(1) Let us assume that the elliptic part D_1 of mixed domain D coincides with the upper semi-strip bounded by the straight lines $x = 0$ $x = 1$ and the solution $u(x, y)$ of problem T for equation (4.4), with zero boundary conditions on the elliptic part of the boundary of domain D is required, which converges to zero for $y \to \infty$, $0 \leqslant x \leqslant 1$, and assumes given values $\psi(x)$ on characteristic AC [cf. G. V. Rudniev (1)].

In this case, by (4.63) the construction of the solution for problem T by means of elementary conformal transformations leads to the problem of finding the function $F(z) = u(x, y) + iv(x, y)$, which is holomorphic in the upper semi-plane $\operatorname{Im} z > 0$ converges to a finite limit for $z \to \infty$, continuous right up to the boundary and satisfies the conditions:

$$\left.\begin{aligned} u(x, 0) &= 0, \quad -\infty < x \leqslant 0, \quad 1 \leqslant x < \infty, \\ u(x, 0) + v(x, 0) &= \varphi(x), \quad 0 \leqslant x \leqslant 1, \quad \varphi(0) = 0. \end{aligned}\right\} \tag{4.88}$$

By (4.88) we have for the function

$$\Phi(z) = \frac{(1-i)F(z)}{\sqrt[4]{(z^3(1-z))}};$$

where by $\sqrt{(z^3(1-z))}$ the single valued branch of this function in the plane cut along $(-\infty, 0)$, $(1, \infty)$ is understood which is positive in $0 < z < 1$, the boundary conditions:

$$\operatorname{Re} \Phi(x) = \begin{cases} \dfrac{\varphi(x)}{\sqrt[4]{(x^3(1-x))}}, & 0 < x < 1, \\ 0, \quad -\infty < x < 0, & 1 < x < \infty. \end{cases}$$

$$\operatorname{Im} \Phi(\infty) = 0.$$

Making use of (4.56) for determining function $F(z)$, we get [cf. A. B. Bitsadze (7)]

$$F(z) = \frac{1}{\pi i(1-i)} \int_0^1 \sqrt{\left(\frac{z^3(1-z)}{t^3(1-t)}\right)} \frac{\varphi(t)\,dt}{t-z}.$$

The rest of the computations leading to an explicit expression for the desired solution do not present any difficulties.

(2) Let D be the mixed domain bounded by the curve σ and characteristics AC and BC of equation (4.4). Denote by

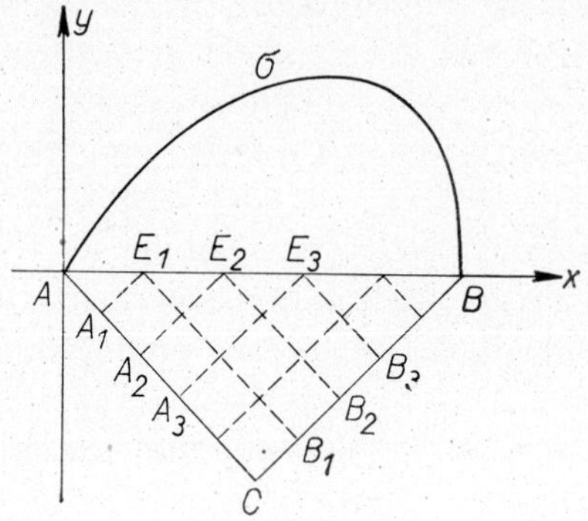

FIG. 7

$E_k(a_k, 0)$, $k = 1, \ldots, n$, $0 < a_1 < a_2 < \ldots < a_n < 1$ given points on the segment AB. Evidently $A_k(1/2\, a_k, -1/2\, a_k)$, $B_k(1/2\, a_k + 1/2, \ 1/2\, a_k - 1/2)$ $k = 0, \ldots, n+1$ $(a_0 = 0, a_{n+1} = 1)$ lie on characteristics AC and BC respectively (cf. Fig. 7).

Problem T_1. Find a function $u(x, y)$ satisfying the conditions: (1) $u(x, y)$ is a solution of equation (4.4) everywhere in domain D apart from points of the segment AB of the real axis and of the characteristics E_kA_k, E_kB_k; (2) it is continuous in the closed domain $\overline{D}$, and the partial derivatives $\partial u/\partial x$ and $\partial u/\partial y$ are continuous at all points of segment AB, apart from perhaps points $A, E_1, \ldots, E_n, B$, where they may become infinite at rates not faster than of order unity; (3) it assumes the given values

$$u = \varphi \text{ on } \sigma, \tag{4.89}$$

$$u = \psi_k \text{ on } A_k A_{k+1} \text{ for even } k, \tag{4.90}$$

$$u = \psi_k \text{ on } B_k B_{k+1} \text{ for odd } k. \tag{4.91}$$

Note that for $n = 0$ condition (4.91) falls away and (4.90) becomes the single condition: $u = \psi_0 (x)$ on AC. Therefore problem T_1 is a direct generalization of problem T.

Problem T_1 was first stated and studied for $n = 1$ in the paper of Gellerstedt (3) in the case of equation (4.3).

In the investigation of problem T_1 no fundamental difficulties arise in comparison with problem T.

From (4.90) and (4.91) we find the equations

$$\frac{\partial u}{\partial x} - \frac{\partial u}{\partial y} = 2 \frac{\mathrm{d}}{\mathrm{d}x} \psi_{2k}\left(\frac{x}{2}\right), \; y = 0, a_{2k} < x < a_{2k+1}, \tag{4.92}$$

$$\frac{\partial u}{\partial x} + \frac{\partial u}{\partial y} = 2 \frac{\mathrm{d}}{\mathrm{d}x} \psi_{2k-1}\left(\frac{1+x}{2}\right), \quad a_{2k-1} < x < a_{2k}. \tag{4.93}$$

When $\psi_k = 0$, $k = 0, 1, \ldots, n$, we see from (4.92) and (4.93), as in § 2, that solution $u(x, y)$ of problem T cannot assume its non-zero extremum for domain $\bar{D}$ in the intervals $a_k < x < a_{k+1}$ $(k = 0, 1, \ldots, n)$ of segment AB. Solution $u(x, y)$ cannot have a non-zero extremum at points $E_k (a_k, 0)$ either. Indeed, we note the following simple property possessed by the solution $u(x, y)$ of the vibrating-string equation

$$u(M_1) + u(M_3) = u(M_2) + u(M_4), \tag{4.94}$$

where M_1, M_2, M_3, M_4 are consecutive vertices of the characteristic rectangle.

When n is even we conclude from (4.94) that $u(a_k, 0) = 0$, $k = 1, 2, \ldots, n$.

Consider now the case when n is not even. By (4.94) it follows that $u(a_1, 0) = \ldots = u(a_n, 0)$. Assume that a points E_k function $u(x, y)$ assumes a non-zero extremum. Separate one of these points E_k from the elliptic part D_1 of domain D by the level curve Γ: $u(x, y) = \text{const}$ with the endpoints lying on the segment AB and contained entirely in domain D_1. Apply to the domain bounded by curve Γ and an interval of the real axis the Green's formula:

$$\iint\left[\left(\frac{\partial u}{\partial x}\right)^2+\left(\frac{\partial u}{\partial y}\right)^2\right]\mathrm{d}x\,\mathrm{d}y=-\int(u-\text{const})\frac{\partial u}{\partial N}\,\mathrm{d}s,$$

where N is the inner normal. We conclude from this formula, and the equations $\partial u/\partial x - \partial u/\partial y = 0$, $a_{2k} < x < a_{2k+1}$, $\partial u/\partial x + \partial u/dy = 0$, $a_{2k-1} < x < a_{2k}$, that $u\,(x,\ y) = \text{const}$ everywhere in domain D_1, and this is impossible whenever $\varphi \neq 0$.

In this way the solution for problem T_1 with zero initial conditions (4.90) and (4.91) assumes its non-zero extremum for the closed domain $\bar{D}_1$ on curve σ (this is the extremal principle for problem T_1). From this principle it follows straight away that problem T_1 cannot have more than one solution.

We now proceed to prove the existence of the solution for problem T_1.

We note that just as in the case of problem T, conditions (4.89) can be replaced without loss of generality by the homogeneous condition

$$u = 0 \text{ on } \sigma. \tag{4.95}$$

The additional assumptions that the curve σ is smooth and that it satisfies the Liapunov condition and that $\partial u/\partial x$ and $\partial u/\partial y$ are continuous everywhere in the closed domain $\bar{D}_1$ apart from perhaps at points A, E_1, . . . , E_n, B will be made.

Denote by $\Phi\,(z)$ the function $u\,(x,\ y) + iv\,(x,\ y)$, which is holomorphic in domain D_1 and satisfies $\Phi\,(0) = 0$.

On the basis of the Cauchy-Riemann conditions, we get from (4.92) and (4.93) that

$$\left.\begin{aligned}&\operatorname{Re}(1-i)\,\Phi\,(x)=2\psi_{2k}\left(\frac{x}{2}\right)+c_{2k},\ a_{2k}\leqslant x\leqslant a_{2k+1},\\&\operatorname{Im}(1-i)\,\Phi\,(x)=-\,2\psi_{2k-1}\left(\frac{x+1}{2}\right)+\\&\qquad\qquad+c_{2k-1},\ b_{2k-1}\leqslant x\leqslant a_{2k},\end{aligned}\right\}\tag{4.96}$$

where $c_0 = 0$, and c_k are arbitrary constants.

In this way the solution of problem T_1 reduces to the determination of function $\Phi(z)$ as a holomorphic function in domain D_1 from the boundary conditions (4.95) and (4.96). In the same way as in the preceding paragraph this can be done by means of a conformal mapping when σ coincides with the semicircle σ_0.

Thus, it will be assumed that σ coincides with σ_0. In this case we conclude from (4.95) that $\Phi(z)$ can be analytically continued to the whole of the upper semi-plane and

$$\left.\begin{aligned} &\operatorname{Im}(1-i)\,\Phi(x) = 2\psi_{2k}\left(\frac{1}{2}\,\frac{x}{2x-1}\right) + c_{2k},\\ &-\infty < x \leqslant a_{2j},\ b_{2j+1} \leqslant x < \infty,\ b_{2k+1} \leqslant x \leqslant b_{2k},\\ &\operatorname{Re}(1-i)\,\Phi(x) = -2\psi_{2k-1}\left(\frac{1}{2}\,\frac{x}{2x-1}\right) + c_{2k-1},\\ &\qquad b_{2k} \leqslant x \leqslant b_{2k-1}, \end{aligned}\right\} \tag{4.97}$$

where it is assumed that $a_{2j} < 1/2 < a_{2j+1}$, $b_k = \dfrac{a_k}{2a_k - 1}$ (the solution corresponding to $a_{2j} = 1/2$ or $a_{2j+1} = 1/2$, is obtained from the formula to be proved below by passing to the limit).

The function $\Phi(z)$, which satisfies (4.96) and (4.97) and is bounded at infinity and near the endpoints a_{2k}, b_{2k}, can be calculated directly [cf. A. V. Bitsadze (6)]:

$$\begin{aligned} (1-i)\,\Phi(z) &= \frac{1}{\pi i}\,\frac{R_1(z)}{R_2(z)}\times\\ &\times\sum_{k=0}\int_{a_{2k}}^{a_{2k+1}}\frac{R_2(t)}{R_1(t)}\left(\frac{1}{t-z}+\frac{1}{(2t-1)(t+z-2tz)}\right)\left[2\psi_{2k}\left(\frac{t}{2}\right)+\right.\\ &\left.+\,c_{2k}\right]dt - \frac{1}{\pi}\,\frac{R_1(z)}{R_2(t)}\sum_{k=1}\int_{a_{2k-1}}^{a_{2k}}\frac{R_2(t)}{R_1(t)}\times\\ &\times\left(\frac{1}{t-z}-\frac{1}{(2t-1)(t+z-2tz)}\right)\times\\ &\times\left[2\psi_{2k-1}\left(\frac{t+1}{2}\right)-c_{2k-1}\right]dt + c\,\frac{R_1(z)}{R_2(z)}, \end{aligned} \tag{4.98}$$

where c is an arbitrary constant, and $R_1(z)$ and $R_2(z)$, for example, for $n = 2m$, are furnished by the formulae:

$$R_1(z) = \sqrt{\left(z \prod_1^m (z - a_{2k})(z - b_{2k})\right)},$$

$$R_2(z) = \sqrt{\left((z-1) \prod_1^m (z - a_{2k-1})(z - b_{2k-1})\right)},$$

$R_1(z)/R_2(z)$ meaning the branch of this function which is holomorphic in the plane cut along $(a_{2k}, a_{2k+1}), (b_{2k}, b_{2k-1})$ and assumes the value 1 at the infinity. Due to the uniqueness of the solution for problem T_1 the constants c and c_k contained in (4.98) can always be chosen in such a way that $u(x, y)$ is bounded in the neighbourhood of points $z = a_{2k-1}$, $z = b_{2k-1}$.

After $u(x, y)$ is found in domain D_1, its calculation in the hyperbolic part of domain D does not present any difficulties.

As in the case of problem T, the extremal principle makes it possible to prove the existence of the solution of problem T_1 in the general case, i. e. without the above made restrictions concerning the partial derivatives of the solution in the closed domain $\bar{D}$ and the smoothness of curve σ.

3. Let D be the doubly-connected mixed domain bounded by the simple curves σ_1 and σ_2, lying in the elliptic half-plane and by the characteristics AC, CB_1, BC_1 and A_1C_1 of equation (4.4) (cf. Fig. 8).

No difficulty is presented by the following mixed problem called problem T_2 in our paper (6). Determine a function $u(x, y)$ with the following properties:

(1) $u(x, y)$ is the solution of equation (4.4) in domain D for $y \neq 0$; (2) it is continuous in closed domain $\bar{D}$, and the partial derivatives $\partial u/\partial x$ and $\partial u/\partial y$ are continuous at all points of the open intervals AB and A_1B_1, but at the points A, B, A_1, B_1 they may become infinite of order lower than unity; (3) it satisfies the boundary conditions

$$u|_{\sigma_1} = \varphi_1, \; u|_{\sigma_2} = \varphi_2, \; u|_{AC} = \psi_1, \; u|_{A_1C_1} = \psi_2,$$

where φ_1, φ_2, ψ_1, ψ_2 are given functions.

4. The problem T for the linear equation of mixed type with main part $k(y)\, \partial^2 u/\partial x^2 + \partial^2 u/\partial y^2$, where $k(y) \gtreqless 0$ for

$y \gtrless 0$, was investigated in the doctoral thesis of K. I. Babenko (1). Specifically, the uniqueness of the solution of this problem for the Chaplygin equation $k(y)\, \partial^2 u/\partial x^2 + \partial^2 u/\partial y^2 = 0$ was the object of study for several authors [cf. Agmon Nirenberg, Protter (1), Protter (1), U Sin Mo. Din Sia Si (1)].

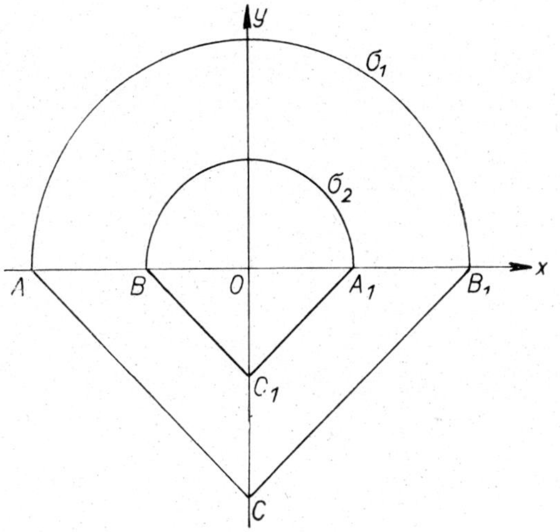

FIG. 8

Problem T is quite well worked out for the equation of the type

$$y^m \frac{\partial^2 u}{\partial x^2} + \frac{\partial^2 u}{\partial y^2} - cu = F(x, y), \quad c = \text{const} > 0 \qquad (4.99)$$

in the paper of Gellerstedt (1) [cf. also Kapilevich (1)].

The structural properties of the solutions of equation (4.2) are studied in the papers of Bergman (1), Germain (1), and L. V. Ovsianikov (1).

5. Undoubtedly, the study of systems consisting of two partial differential equations of mixed type is of scientific interest. Apparently we have first to investigate the generalization of problem T for a system of the type:

$$y^{2m+1} \frac{\partial u}{\partial x} - \frac{\partial v}{\partial y} = au + bv, \qquad \frac{\partial u}{\partial y} + \frac{\partial v}{\partial x} = cu + dv \quad (4.100)$$

in the following form: in the mixed domain D, bounded by the Jordan curve σ, in the elliptic part of the half-plane, and by characteristics AC and CB of system (4.100) in the hyperbolic part of the half-plane, find solutions $u(x, y)$ and $v(x, y)$ of this system satisfying the boundary conditions

$$au + \beta v = \lambda \text{ on } \sigma w \text{ on } AC, \qquad (4.101)$$

where a, β and γ are given functions.

In a special case a problem of this kind was studied in the paper of Friedrichs (1). The paper of Z. A. Kikvidze also must be noted in which for the system

$$\frac{\partial u}{\partial x} - \frac{\partial v}{\partial y} = 0, \qquad \frac{\partial u}{\partial y} + \operatorname{sgn} y \cdot \frac{\partial v}{\partial x} = 0$$

the problem of the type of (4.101) was investigated under extremely special conditions concerning a and β.

6. To our view the study of higher order equations and systems of equations of mixed type is not less important. As a model equation of the fourth order of mixed type we could conveniently take for instance the equation

$$\frac{\partial^4 u}{\partial x^4} + 2\operatorname{sgn} y \cdot \frac{\partial^4 u}{\partial x^2 \partial y^2} + \frac{\partial^4 u}{\partial y^4} = 0 \qquad (4.102)$$

or the equation

$$\left(y \frac{\partial^2}{\partial x^2} + \frac{\partial^2}{\partial y^2}\right)\left(y \frac{\partial^2 u}{\partial x^2} + \frac{\partial^2 u}{\partial y^2}\right) = 0.$$

One mixed problem for (4.102) has been studied in the paper of M. M. Smirnov (1).

7. The difficulty of the correct statement of the problem for equations of mixed type in higher dimensions still remains.

Consider the equation of mixed type

$$\Delta u + \operatorname{sgn} t \cdot \frac{\partial^2 u}{\partial t^2} = 0, \qquad (4.103)$$

where Δ is the Laplace operator for the spatial variables x_1, $x_2, \ldots, x_n$.

Denote by D the mixed domain in the variables $x_1, \ldots, x_n, t$, which is bounded by the semi-sphere σ: $r^2 + t^2 = 1$, $t > 0$ and the characteristic cones:

$$K_1\colon\ t = r - 1,\ -\frac{1}{2} \leqslant t \leqslant 0,$$

$$K_2\colon\ t + r = 0,\ -\frac{1}{2} \leqslant t \leqslant 0,$$

where $r^2 = x_1^2 + \ldots + x_n^2$.

The simplest mixed problem for equation (4.103) can be constructed in the following manner.

We try to determine a function $u(x_1, \ldots, x_n, t)$ with the properties: (1) u is the solution of equation (4.103) in domain D for $t \neq 0$; (2) it is continuous in the closed domain $\overline{D}$; (3) all partial derivatives of the first order of u are continuous in D right up to σ, but on the manifolds $r = 1$, $t = 0$; $r = \frac{1}{2}$, $t = \frac{1}{2}$ and at the origin of the co-ordinate system they may have weak singularities in the sense that they are integrable on the hypersurface boundary of domain D; (4) on σ and on K_1 the function u assumes the given values:

$$u = \varphi \text{ on } \sigma,\ u = \psi \text{ on } K_1. \tag{4.104}$$

In the case when the vertex of cone K_2 is removed from the centre of coordinates into domain $r \leqslant 1$, $t = 0$ (Protter), the above formulation of this problem may prove to be incorrect.

The proof of the existence of a solution of equation (4.103) may be carried out by means of the inversion of multi-dimensional singular integral equations of the first kind, which are the spatial analogues of the familiar Abel and Tricomi equations. In a special case this problem has been studied by the present author (8).

CHAPTER 5

OTHER MIXED PROBLEMS

§ 1. The Mixed Problem M

Assume that the boundary of the mixed domain D is a contour of the third kind in the terminology of Tricomi, i.e. that in the hyperbolic half-plane a piece of characteristic AC is replaced by a piece L of some curve which satisfies certain conditions to be described later.

The boundary conditions of mixed problem M are given for σ and for L. F. I. Frankl (2, 3) directed attention to the hydrodynamic meaning of this problem. For the sake of simplicity we shall in considering this problem restrict ourselves to the case of the Lavrent'ev equation (4.4).

Let D be a singly-connected domain bounded by (a) the Jordan curve σ with endpoints at A (0, 0) and B (1, 0), situated in the upper semi-plane $y > 0$; (b) the monotone curve L: $y = -\gamma(x)$, $0 \leqslant x \leqslant l$, situated inside the characteristic triangle ACB, $\gamma(0) = 0$, $l + \gamma(l) = 1$ and (c) characteristic C_1B: $y = x - 1$, $l \leqslant x \leqslant 1$ of equation (4.4) (Fig. 9).

Problem M consists of determining a function $u(x, y)$ possessing the following properties: (1) $u(x, y)$ is the solution of equation (4.4) in domain D for $y \neq 0$; (2) it is continuous in the closed domain $\overline{D}$ and its partial derivatives $\partial u/\partial x$ and $\partial u/\partial y$, are continuous inside the domain D; (3) on the curves σ and L it assumes the prescribed values

$$u|_\sigma = \varphi, \tag{5.1}$$

$$u|_L = \psi. \tag{5.2}$$

In trying to prove the existence and the uniqueness of a solution for problem M we encounter difficulties caused by the fact, shown below, that the fundamental functional relation

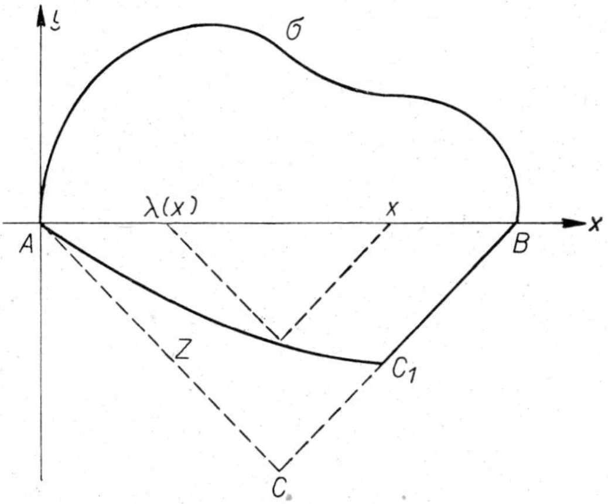

FIG. 9

between $\tau(x) = u(x, 0)$ and $\nu(x) = \partial u/\partial y\,|_{y=0}$, which is generated in the hyperbolic part D_2 of domain D, is of a more complicated nature than relation (4.11).

§ 2. The Proof of the Uniqueness of Solution for Problem M

Let σ be a smooth Jordan curve satisfying the condition

$$y'_s(x - x^2 - y^2) - yx'_s \geqslant 0, \tag{5.3}$$

where s is the length of curve σ, calculated from point B, and $x = x(s)$, $y = y(s)$ are the parametric equations of this curve.

Concerning curve L we assume in addition that its curvature satisfies Hölder's condition, and

$$0 < \frac{d\gamma}{dx} \leqslant 1. \tag{5.4}$$

$$\frac{d\gamma}{dx} \leqslant \frac{\gamma}{x - x^2 + \gamma^2}. \tag{5.5}$$

It will also be assumed that $\partial u/\partial x$ and $\partial u/\partial y$ are continuous in the closed $\overline{D}$ domain except, perhaps, at points A and B, where they may become infinite of order not higher than unity. In addition the function is continuously differentiable once, $\psi(x)$ twice and $\psi''(x)$ satisfies Hölder's condition.

Under these assumptions we prove the uniqueness of the solution of the homogeneous problem

$$u|_{\sigma} = 0 \tag{5.1$_0$}$$

$$u|_{L} = 0. \tag{5.2$_0$}$$

We make the preliminary remark that the general solution of equation (4.4) satisfying condition (5.2) in domain D_2 can be represented in the form

$$u(x, y) = f(x+y) - f\{\delta(x-y) - \gamma[\delta(x-y)]\} + \\ + \psi[\delta(x-y)], \tag{5.6}$$

where $x = \delta(\xi)$, $0 \leqslant \xi \leqslant$ is a known function determined uniquely from the equation

$$x + \gamma(x) = \xi, \quad 0 \leqslant x \leqslant l, \tag{5.7}$$

and $f(t)$ is an arbitrary twice continuously differentiable function in $0 < t < 1$, which is continuous in the closed interval $0 \leqslant t \leqslant 1$ which a possible, integrable, singularity occurring at the endpoints of this interval for $f'(t)$.

Indeed, take the general solution of equation (4.4) in domain D_2 in the form

$$u(x, y) = f(x+y) + f_1(x-y). \tag{5.8}$$

According to (5.2) and (5.8) we have

$$f[x - \gamma(x)] + f_1[x + \gamma(x)] = \psi(x), \;\; 0 \leqslant x \leqslant l.$$

Hence, taking notation (5.7) into account, we get

$$f_1(\xi) = \psi[\delta(\xi)] - f\{\delta(\xi) - \gamma[\eta(\xi)]\}, \;\; 0 \leqslant \xi \leqslant 1.$$

Substituting the found expression for $f(\xi)$ into (5.8) we see that (5.6) is correct.

In particular, according to (5.6)

$$\tau'(x) + \nu(x) = 2f'(x). \tag{5.9}$$

Formula (5.9) shows that $f'(x)$ can become infinite for $x \to 0$ and for $x \to 1$ not faster than of order unity.

Under condition (5.2_0), the representation (5.6) becomes

$$u(x, y) = f(x + y) - f\{\delta(x - y) - \gamma[\delta(x - y)]\}. \quad (5.6_0)$$

Hence it is clear that $f(0)$ can be taken to be zero without restricting generality.

From formula (5.6_0) we get

$$u(x, 0) = f(x) - f\{\delta(x) - \gamma[\delta(x)]\}. \quad (5.10)$$

Denote by $F(z)$ the function $u(x, y) + iv(x, y)$ which is analytic in the elliptic part D_1 of domain D. Since $u(0, 0) = 0$, function $F(z)$ can be made to satisfy the condition:

$$F(0) = 0. \quad (5.11)$$

Due to the equations

$$\left(\frac{\partial u}{\partial y}\right)_{y=+0} = \left(\frac{\partial u}{\partial y}\right)_{y=-0}, \quad \frac{\partial v}{\partial x} = -\frac{\partial u}{\partial y}, \quad (x, y) \in D_1,$$

we get from (5.6)

$$\frac{dv(x, 0)}{dx} = -\frac{d}{dx} f(x) - \frac{d}{dx} f\{\delta(x) - \gamma[\delta(x)]\}.$$

Hence, after integrating, due to (5.11), we have

$$v(x, 0) = -f(x) - f\{\delta(x) - \gamma[\delta(x)]\}. \quad (5.12)$$

Consider the integral

$$J = -\int_0^1 u(x, 0)\, v(x, 0) \frac{1 - x}{x} dx =$$
$$= \int_0^1 \{f^2(x) - f^2[\delta(x) - \gamma(\delta(x)]\} \frac{1 - x}{x} dx, \quad (5.13)$$

the existence of which is known.

After a transformation of the variables of integration the expression (5.13) assumes the form

$$J = \int_{2l-1}^{1} f^2(x) \frac{1-x}{x} \mathrm{d}x + \int_{0}^{l} f^2 [x - \gamma(x)] \Big[\frac{1 - x + \gamma(x)}{x - \gamma(x)} - $$

$$- \frac{1 + \gamma'(x)}{1 - \gamma'(x)} \, \frac{1 - x - \gamma(x)}{x + \gamma(x)} \Big] [1 - \gamma'(x)] \, \mathrm{d}x,$$

whence, by (5.5), we conclude that

$$J \geqslant 0. \tag{5.14}$$

On the other hand, by (5.1_0), we have

$$\int_{\sigma} F^2(z) \frac{1-z}{z} \mathrm{d}z + \int_{0}^{1} [u(x, 0) + iv(x, 0)]^2 \frac{1-x}{x} \mathrm{d}x = 0. \tag{5.15}$$

Separating the imaginary part of (5.15), we get

$$J = - \int_{0}^{1} u(x, 0)\, v(x, 0) \frac{1-x}{x} \mathrm{d}x =$$

$$= - \frac{1}{2} \int_{\sigma} \frac{y_s' (x - x^2 - y^2) - x_s' y}{x^2 + y^2} v^2 \, \mathrm{d}s. \tag{5.16}$$

From (5.16) and (5.3), we conclude that

$$J \leqslant 0. \tag{5.17}$$

On comparing (5.14) and (5.17), we get $f(x) = 0$, $0 \leqslant x \leqslant 1$.

Hence, in turn it follows that $u(x, y)$ vanishes identically in the whole of the domain D.

In this way we find that if σ and L satisfy conditions (5.3), (5.4), (5.5) problem M cannot have more than one solution.

Condition (5.5) is observed, for example, if the curve L is bent inward with respect to the x-axis and condition (5.3) is observed partly, if curve σ is bent inward in relation to the x axis and is situated inside the unit circle $|z| < 1$.

§ 3. Concerning the existence of the solution of problem M

From formula (5.6) we get directly the first functional relation between $\tau(x)$ and $\nu(x)$, which is generated inside the hyperbolic part D_2 of domain D:

$$\tau'(x) - \nu(x) = 2\frac{\mathrm{d}}{\mathrm{d}x}\psi[\delta(x)] -$$

$$- (\tau'\{\delta(x) - \gamma[\delta(x)]\} + \nu\{\delta(x) - \gamma[\beta(x)]\}) \times \qquad (5.18)$$

$$\times \frac{\mathrm{d}}{\mathrm{d}x}\{\delta(x) - \gamma[\delta(x)]\}.$$

The second functional relation between $\tau(x)$ and $\nu(x)$, as was already pointed out, is of the form of (4.25).

In the present paragraph we shall treat the question concerning the existence of the solution of problem M in the case when curve L first coincides with piece AF of characteristic AC, and then deviates from it bending inside the characteristic triangle ACB [cf. A. V. Bitsadze (4—6)].

Denote by $E(h, 0)$ a point on segment AB, $0 < h < 1$. The characteristic of equation (4.4) which starts from point E, intersects characteristics AC and BC in $F(h/2, -h/2)$, $G[(h+1)/2, (h-1)/2\varkappa]$ respectively. Let $H(1-l, -l)$, $(1-h)/2 < l < 1/2$ be a point on characteristic BC. Join points F and H by the curve $y = -\gamma(x)$, $h/2 \leqslant \varkappa \leqslant 1 - l$, where $\gamma''(x)$ satisfies Hölder's condition and also conditions (5.4) and (5.5).

Let L coincide with the curve

$$y = -x, \quad 0 \leqslant x \leqslant \frac{h}{2},$$

$$y = -\gamma(x); \ \frac{h}{2} \leqslant x \leqslant 1 - l$$

and let its curvature satisfy the Hölder condition.

As regards the curve σ we shall assume that it satisfies the Liapunov condition, terminates in some small curves AA' and BB' of the semi-circle σ_0 and, besides that, it is subject to condition (5.3).

Let D be the mixed domain bounded by the curves σ, L and HB (Fig. 10).

By what has already been proved the mixed problem M cannot have more than one solution in D.

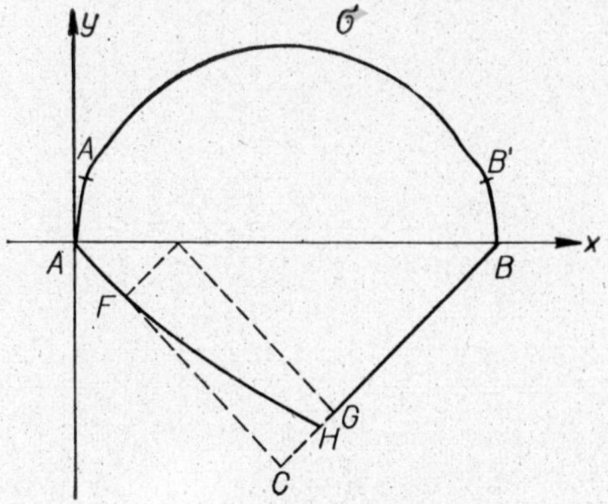

FIG. 10

We shall now prove that in the case considered problem M does have a solution.

Note beforehand that, without loss of generality, it can be assumed that $\varphi = 0$, $\psi(0) = 0$. In what follows we shall make the additional demand that

$$\psi'(0) = 0. \tag{5.19}$$

Consider first the case when σ coincides with the semi-circle σ_0. In this case relation (4.25) becomes of the form of (4.27) and we have from formula (5.18)

$$\tau'(x) - \nu(x) = 2\frac{\mathrm{d}}{\mathrm{d}x}\psi\left(\frac{x}{2}\right), \quad 0 < x \leqslant h, \tag{5.20}$$

$$\tau'(x) - \nu(x) =$$

$$= 2\nu - [\lambda(x)]\frac{\mathrm{d}}{\mathrm{d}x}\lambda(x) - 2\frac{\mathrm{d}}{\mathrm{d}x}\psi\left[\frac{\lambda(x)}{2}\right] + 2\frac{\mathrm{d}}{\mathrm{d}x}\psi[\delta(x)],$$

$$h \leqslant x \leqslant 1, \; 0 \leqslant \lambda(x) = \delta(x) - \gamma[\delta(x)] \leqslant 1 - 2l. \tag{5.21}$$

Eliminating $\tau'(x)$ from (4.27), (5.20) and (5.21) we get

$$\nu(x)+\frac{1}{\pi}\int_0^1\left(\frac{1}{t-x}+\frac{1-2t}{t+x-2tx}\right)\nu(t)\,\mathrm{d}t=F(x), \quad (5.22)$$

where

$$F(x)=\begin{cases}F_0(x), & 0<x<h,\\ 2\nu[\lambda(x)]\dfrac{\mathrm{d}}{\mathrm{d}x}\lambda(x)+E_0(x), & h\leqslant x\leqslant 1,\end{cases}$$

$$F_0(x)=\begin{cases}-2\dfrac{\mathrm{d}}{\mathrm{d}x}\psi\left(\dfrac{x}{2}\right), & 0\leqslant x\leqslant 1,\\ 2\dfrac{\mathrm{d}}{\mathrm{d}x}\psi\left[\dfrac{\lambda(x)}{2}\right]-2\dfrac{\mathrm{d}}{\mathrm{d}x}\psi[\delta(x)], & h\leqslant x<1.\end{cases}$$

Applying the transformation formula (4.60) equation (5.22) can be rewritten in the equivalent form:

$$\nu(x)+\frac{1}{\pi}\int_h^1\sqrt{\left(\frac{x(1-t)}{t(1-x)}\right)}\left(\frac{1}{t-x}+\right.$$
$$\left.+\frac{1-2t}{t+x\,2tx)}\right)\lambda'(t)\,\nu[\lambda(t)]\,\mathrm{d}t=F_1(x) \quad (5.23)$$
$$0<x\leqslant h,$$

$$\nu(x)=F_1(x)+\nu[\lambda(x)]\,\lambda'(x)-$$
$$-\frac{1}{\pi}\int_h^1\sqrt{\left(\frac{x(1-t)}{t(1-x)}\right)}\left(\frac{1}{t-x}+\right.$$
$$\left.+\frac{1-2t}{t=x-2tx}\right)\cdot\lambda'(t)\,\nu[\lambda(t)]\,\mathrm{d}t, \quad (5.24)$$
$$h\leqslant x<1,$$

where

$$F_1(x)=\frac{1}{2}F_0(x)-\frac{1}{2\pi}\int_0^1\sqrt{\left(\frac{x(1-t)}{t(1-x)}\right)}\left(\frac{1}{t-x}+\right.$$
$$\left.+\frac{1-2t}{t+x-2tx}\right)F_0(t)\,\mathrm{d}t.$$

From the foregoing consideration it is obvious that problem M and equations (5.23) and (5.24) are completely equivalent.

With the transformation of variable $\xi = \lambda(t)$, $h \leqslant t \leqslant 1$, $0 \leqslant \xi \leqslant 1 - 2l$ equation (5.23) becomes

$$\nu(x) + \int_0^{1-2l} \sqrt{\left(\frac{x(1-\omega(\xi))}{\omega(\xi)(1-x)}\right)} \left(\frac{1}{\omega(\xi) - x} + \right.$$

$$\left. + \frac{1 - 2\omega(\xi)}{\omega(\xi) + x - 2x\omega(\xi}\right)\nu(\xi)\,\mathrm{d}\xi = F_1(x), \tag{5.25}$$

where ω is the function inverse to λ.

For $x \in (0,1 - 2l)$, due to the inequality $1 - 2l < h$, we conclude that equation (5.25) is a Fredholm integral equation of second kind the solvability of which is a consequence of the uniqueness of the solution for problem M.

Denoting by $R(x, t)$, $x, t \in (0,1 - 2l)$ the resolvent of equation (5.25), the solution of this equation can then be represented by the formula

$$\nu(x) = F_1(x) + \int_0^{1-2l} R(x, t)\, F_1(t)\,\mathrm{d}t, \quad x \in (0,\ 1 - 2l). \tag{5.26}$$

Substituting equation (5.26) for $\nu(x)$, $x \in (0,\ 1 - 2l)$ into formula (5.25), we get

$$\nu(x) = F_1(x) - \frac{1}{\pi} \int_0^{1-2l} \left\{ \sqrt{\left(\frac{x(1-\omega(t))}{\omega(t)(1-x)}\right)} \left[\frac{1}{\omega(t) - x} + \right.\right.$$

$$\left. + \frac{1 - 2\omega(t)}{\omega(t) + x - 2x\,\omega(t)}\right] +$$

$$+ \int_0^{1-2l} R(\xi, t) \sqrt{\left(\frac{x(1-\omega(\xi))}{\omega(\xi)(1-x)}\right)} \left[\frac{1}{\omega(\xi) - x} + \right. \tag{5.27}$$

$$\left.\left. + \frac{1 - 2\omega(\xi)}{\omega(\xi) + x - 2x\omega}\right]\mathrm{d}\xi \right\} F_1(t)\,\mathrm{d}t,$$

$$x \in (1 - 2l, h).$$

After $\nu[\lambda(x)]$, $x \in (h, 1)$, has been calculated, we get from (5.24) the expression of $\nu(x)$ for $x \in (h, 1)$:

$$\nu(x) = F_1(x) + F_1[\gamma(x)]\gamma'(x) + \int_0^{1-2l} \Big\{R[\lambda(x), t]\gamma'(x) - $$
$$- \frac{1}{\pi}\sqrt{\left(\frac{x(1-\omega(t)}{\omega(t)(1-x)}\right)}\left[\frac{1}{\omega(t)-x} + \frac{1-2\omega(t)}{\omega(t)+x-2x\omega(t)}\right] - $$
$$- \frac{1}{\pi}\int_0^{1-2l} R(\xi, t)\sqrt{\left(\frac{x(1-\omega(\xi)}{\omega(\xi)(1-x)}\right)}\left[\frac{1}{\omega(\xi)-x} + \right. \tag{5.28}$$
$$\left. + \frac{1-2\omega(\xi)}{\omega(\xi)+x-2x\omega(\xi)}\right] \times d\xi\Big\} F_1(t)\,dt.$$

From (5.26), (5.27) and (5.28) we get finally

$$\nu(x) = F^*(x)\int_0^{1-2l} R^*(x, t)F_1(t)\,dt, \tag{5.29}$$

where

$$F^*(x) = \begin{cases} F_1(x), x \in (0, h) \\ F_1(x) + F_1[\gamma(x)]\lambda'(x), \;\; x \in (h, 1), \end{cases} \tag{5.30}$$

$$R^*(x, t) = R(x, t), \;\; x, t, \in (0,1-2l),$$

$$\left.\begin{aligned}
&R^*(x, t) = -\frac{1}{\pi}\sqrt{\left(\frac{x(1-\omega(t))}{\omega(t)(1-x)}\right)}\left[\frac{1}{\omega(t)-x} + \right. \\
&\left. + \frac{1-2\omega(t)}{\omega(t)+x-2x\omega(t)}\right] - \frac{1}{\pi}\int_0^{1-2l} R(x, t)\sqrt{\left(\frac{x(1-\omega(\xi))}{\omega(\xi)(1-x)}\right)} \times \\
&\times\left[\frac{1}{\omega(t)-x} + \frac{1-1\omega(\xi)}{x+\omega(\xi)-2x\omega(\xi)}\right]d\xi, \\
&\qquad x \in (1-2l, h), \;\; t \in (0, \; 1-2l), \\
&R^*(x, t) = R[\gamma(x), t]\lambda'(x) - \frac{1}{\pi}\sqrt{\left(\frac{x(1-\omega(t))}{\omega(t)(1-x)}\right)} \times \\
&\times \frac{1}{\omega(t)-x} + \frac{1-2\omega(t)}{\omega(t)+x-2x\omega(t)}\Bigg] - \frac{1}{\pi}\int_0^{1-2l} R(\xi,t) \times \\
&\times \sqrt{\left(\frac{x(1-\omega(\xi))}{\omega(\xi)(1-x)}\right)}\left[\frac{1}{\omega(\xi)-x} + \frac{1-2\omega(\xi)}{\omega(\xi)+x-2x\omega(\xi)}\right]d\xi, \\
&\qquad x \in (h, 1), \;\; t \in (0, \; 1-2l).
\end{aligned}\right\} \tag{5.31}$$

In this way we have proved the existence of a solution of the functional equation (5.22). Hence we have also proved the existence of the solution of problem M in the case considered.

Assume now that σ terminates in some small curves AA', BB' of semi-circle σ_0.

After eliminating $\tau'(x)$ from relations (4.25), (5.20) and (5.21) we get

$$\nu(x) + \frac{1}{\pi}\int_0^1 \left(\frac{1}{t-x} + \frac{1-2t}{t+x-2tx}\right)\nu(t)\,dt = F_0(x) - \int_0^1 K(x,t)\,\nu(t)\,dt, \qquad x \in (0,h), \tag{5.32}$$

$$\nu(x) + \frac{1}{\pi}\int_0^1 \left(\frac{1}{t-x} + \frac{1-2t}{t+x-2tx}\right)\nu(t)\,dt - 2\nu[\lambda(x)]\,\lambda'(x) = F_0(x) - \int_0^1 K(x,t)\,\nu(t)\,dt, \quad x \in (h,1), \tag{5.33}$$

where $K(x, t)$ is a regular kernel of the form of (4.49).

For the time being the expression $\int_0^1 K(x, t)\,\nu(t)\,dt$ will be considered as known in equations (5.32) and (5.33) applying (5.39) we can replace these equations by the equivalent equations

$$\nu(x) + \int_0^1 K^{**}(x,t)\nu(t)\,dt = F_{**}(x), \tag{5.34}$$

where

$$F_{**}(x) = \begin{cases} F_1(x) + \int_0^{1-2l} R^*(x,t)\,F_1(t)\,dt, \quad x \in (0,h), \\ F_1(x) + F_1[\lambda(x)]\,\lambda'(x) + \\ \quad + \int_0^{1-2l} R^*(x,t)\,F_1(t)\,dt, \quad x \in (h,1). \end{cases}$$

$$K^{**}(x,t) = \begin{cases} K^*(x,t) + \int\limits_0^{1-2l} R^*(x,\xi) K^*(\xi,t)\, d\xi, \quad x \in (0,h), \\ K^*(x,t) + K^*[\lambda(x),t]\,\lambda'(x) + \\ \quad + \int\limits_0^{1-2l} R^*(x,\xi) K^*(\xi,t)\, d\xi, \\ \quad x \in (h,1), \end{cases}$$

$$K^*(x,t) = \frac{1}{2} K(x,t) -$$

$$- \frac{1}{2\pi} \int\limits_1^0 \sqrt{\left(\frac{x(1-\xi)}{\xi(1-x)}\right)} \left(\frac{1}{t-\xi} + \frac{1-2\xi}{t+\xi-2t\xi}\right) K(\xi,t)\, d\xi.$$

Equation (5.34) is a Fredholm integral equation of the second kind.

Since the equivalence is preserved everywhere, from the uniqueness of the solution of problem M the existence of a solution of the integral equation (5.34) follows. Hence, in turn, the existence of the solution of problem M automatically follows [cf. the work of the present author (4—6)].

The restriction on curve σ, included among the conditions, namely that is should terminate in some small curves AA', BB' of semi-circle σ_0, can be abandoned. However, the restriction (5.3) placed on σ is essential for us in that the proof of the existence of the solution of problem M made use of the uniqueness of this solution.

Whenever the curve L deviates from characteristic AC from its start the existence proof of the solution for problem M becomes more difficult.

To finish we point out a reduction of problem M.

From formula (5.6) we have

$$u(x,0) = f(x) - f[\lambda(x)] + \psi[\delta(x)], \tag{5.35}$$

where

$$\lambda(x) = \delta(x) - \gamma[\delta(x)].$$

On the other hand, making use of the Cauchy-Riemann equation

$$\frac{\partial v}{\partial x} = - \frac{\partial u}{\partial y}$$

and the condition concerning the continuity of $\partial u/\partial x$ and $\partial u/\partial y$ across segment AB, by (5.6), we may write

$$v(x, 0) = -f(x) - f[\lambda(x)] + \psi[\delta(x)], \quad v(0, 0) = 0. \quad (5.36).$$

Eliminating $f(x)$, from relations (5.35) and (5.36), we get

$$u(x, 0) + v(x, 0) + u[\lambda(x), 0] - v[\lambda(x), 0] = 2\psi[\delta(x)],$$
$$0 \leqslant x \leqslant 1.$$

In this way problem M is reduced to an entirely new problem of the theory of functions: find a function $F(z) = u(x, y) + iv(x, y)$, which is holomorphic in domain D_1 continuous in the closed domain $\overline{D}_1$ and satisfies the conditions:

$$\operatorname{Re} F|_\sigma = \varphi, \quad F(0) = 0,$$
$$\operatorname{Re}(1 - i)\{F(x) + iF[\lambda(x)]\} = 2\psi[\delta(x)].$$

In the next chapter we shall prove that the general mixed problem, of which problem M is a special case has a solution. For this reason we shall not carry out here a detailed proof of the existence of the solution of the mixed problem M.

Problem M for equation (4.1) in the case when curve L has the common segment AF with characteristic AC, was considered in the papers of F. I. Frankl (4) and Protter (4).

§ 4. The General Mixed Problem

We shall again take the Lavrent'ev equation as the model equation of mixed type.

Denote by D the singly-connected mixed domain bounded by the smooth Jordan curve σ with endpoints $A(0, 0)$, $B(1, 0)$ lying in the upper half-plane $y > 0$ and by the curves L: $y = -\gamma(x)$, $L_1: y = -\gamma_1(x)$, starting from those points having curvatures satisfying Hölder's condition. In addition we shall demand that $\gamma(x)$ and $\gamma_1(x)$ satisfy the conditions

$$\gamma > 0, \; \gamma_1 > 0, \; 0 < \gamma'(x) < 1, \; 0 < -\gamma_1'(x) < 1. \quad (5.37)$$

Let $C\,[x_1, -\gamma(x_1)]$ be the intersection of curves L and L_1. From points $E(x_0, 0)$, $x_1 - \gamma(x_1) \leqslant x_0 \leqslant x_1 + \gamma(x_1)$, draw characteristics $EB_2 : y = x - x_0$ and $EB_1 : y = x_0 - x$ of equation (4.4) where B_2 and B_1 are the points at the intersection of the said characteristics with curves L and L_1. Denote by L_2 and L_3 curves AB_2 and BB_1 of curves L and L_1 respectively (Fig. 11).

General mixed problem:

find the function $u(x, y)$ with the following properties: (1) $u(x, y)$ is the solution of equation (4.4) in domain D for $y \neq 0$, $y \neq x - x_0$, $y \neq x_0 - x$; (2) it is continuous in the

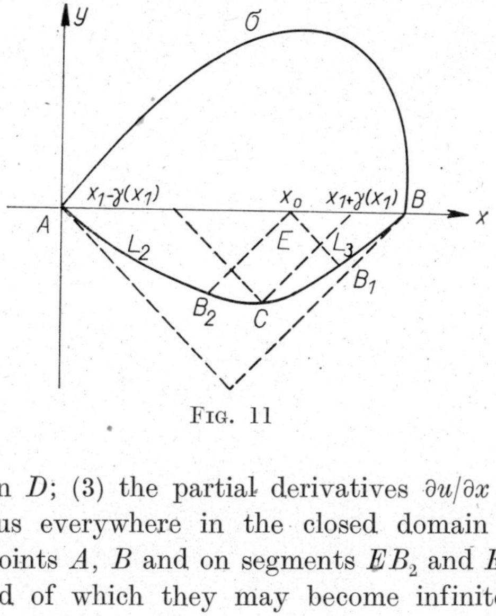

FIG. 11

closed domain D; (3) the partial derivatives $\partial u/\partial x$ and $\partial u/\partial y$ are continuous everywhere in the closed domain $\overline{D}$ except perhaps, at points A, B and on segments EB_2 and EB_1, in the neighbourhood of which they may become infinite of order not higher than 1/2; (4) on σ, L_2 and L_3 function $u(x, y)$ assumes the given values

$$u|_\sigma = \psi_1, \quad u|_{L_2} = \psi_2, \quad u|_{L_3} = \psi_3, \tag{5.38}$$

where ψ_1 is continuous and ψ_2 and ψ_3 are twice differentiable functions such that their second derivatives satisfy Hölder's condition.

It can be assumed without loss of generality that $\psi_1 = 0$, In addition we demand that

$$\psi_2'(0) = \psi_3'(0) = 0. \tag{5.39}$$

For the sake of simplifying the calculations we assume that $\gamma = ax$, $\gamma_1 = -\beta x + \beta$, where a and β are constants such that $0 < a < 1$, $0 < \beta < 1$. In this case

$$C_1 = C_1\left(\frac{\beta}{a+\beta}, \frac{-a\beta}{a+\beta}\right), \quad B_1 = B_1\left(\frac{x_0+\beta}{x_0+\beta}, \frac{\beta(x_0-1)}{1+\beta}\right),$$

$$B_2 = B_2\left(\frac{x_0}{1+a'}, \frac{1+a}{-ax_0}\right).$$

This result remains in force also for the more general assumption made concerning L and L_1, as formulated above [cf. (5.37)].

The general solution of equation (4.4) satisfying (5.38) have the following forms in triangles AB_2E and EB_1B, respectively.

$$u(x, y) = f(x+y) - f[\lambda(x-y)] + + \psi_2\left[\frac{1+\lambda}{2}(x-y)\right], \tag{5.40}$$

$$u(x, y) = \varphi(x-y) - \varphi[\mu(x+y) + 1 - \mu] + + \psi_3\left[\frac{(x+y)(1+\mu) + 1 - \mu}{2}\right], \tag{5.41}$$

where $f(t)$, $t \in (0, x_0)$, $\varphi(t)$, $t \in (x_0, 1)$ are arbitrary twice differentiable functions with possible singularities of order not exceeding $1/2$ possessed by the first derivatives at the endpoints of these intervals and $\lambda = \dfrac{1-a}{1+a}$, $\mu = \dfrac{1-\beta}{1+\beta}$.

We have from (5.40) and (5.41)

$$\frac{\partial u(x, y)}{\partial x} + \frac{\partial u(x, y)}{\partial y} = 2\frac{\mathrm{d}}{\mathrm{d}x} f(x),\ y = 0,\ 0 < x < x_0, \tag{5.42}$$

$$\frac{\partial u(x, y)}{\partial x} - \frac{\partial u(x, y)}{\partial y} = 2\frac{\mathrm{d}}{\mathrm{d}x}\varphi(x),\ \ y = 0,\ x_0 < x < 1, \tag{5.43}$$

$$\frac{\partial u(x\, y)}{\partial x} - \frac{\partial u(x, y)}{\partial y} = 2\frac{\mathrm{d}}{\mathrm{d}x} f(\lambda x) + 2\frac{\mathrm{d}}{\mathrm{d}x}\psi_2\left(\frac{1+\lambda}{2}x\right), \quad y = 0,\ \ 0 < x < x_0, \tag{5.44}$$

$$\frac{\partial u(x,y)}{\partial x}+\frac{\partial u(x,y)}{\partial y}=-2\frac{d}{dx}\varphi(\mu x+1-\mu)+$$
$$+2\frac{d}{dx}\psi_3\left[\frac{x(1+\mu)+1-\mu}{2}\right], \tag{5.45}$$
$$y=0,\ x_0<x<1.$$

We prove first the uniqueness of the solution of the general mixed problem. For this purpose we consider function $F_0(z) = u_0(x, y) + iv_0(x, y)$, which is holomorphic in the elliptic part D_2 of domain D and the real part of which is a solution of the homogeneous general mixed problem ($\psi_2 = \psi_3 = 0$).

From familiar properties of holomorphic functions, we conclude that

$$\int_\sigma z(1-z)(x_0-z)F_0'^2(z)\,dz=$$
$$=-\int_0^1 x(1-x)(x-x_0)F_0'^2(x)\,dx. \tag{5.46}$$

Separating the imaginary part of equation (5.46) and demanding addition from σ that

$$\omega(s)=\operatorname{Im}\left\{z(1-z)(x_0-z)\left(\frac{dy}{dx}+i\right)^2 dz\right\}_\sigma\leqslant 0, \tag{5.47}$$

we have

$$J=\int_0^1 x(1-x)(x_0-x)\frac{\partial u_0}{\partial x}\frac{\partial u_0}{\partial y}\,dx=\int_\sigma\omega(s)\left(\frac{\partial u_0}{\partial y}\right)^2 ds\leqslant 0, \tag{5.48}$$

where s is the length of curve σ, measured from point B.

Note that in particular, whenever σ coincides with the semicircle σ_0, we have $\omega(s) = -\dfrac{1}{8}$.

On the other hand, from (5.42), (5.43), (5.44) and (5.45) we have

$$\frac{\partial u_0}{\partial x}\frac{\partial u_0}{\partial y}=\left[\frac{d}{dx}f(x)\right]^2-\left[\frac{d}{dx}f(x\lambda)\right]^2,\ y=0\ 0<x<x_0$$

$$\frac{\partial u_0}{\partial x}\frac{\partial u_0}{\partial y}=-\left[\frac{d}{dx}\varphi(x)\right]^2+\left[\frac{d}{dx}\varphi(\mu x+1-\mu)\right]^2,$$
$$y=0,\ x_0<x<1.$$

Therefore, we can write

$$J = \int_0^{x_0} x(1-x)(x_0-x)\left\{\left[\frac{d}{dx} f(x)\right]^2 - \left[\frac{d}{dx} f(\lambda x)\right]^2\right\} dx +$$

$$+ \int_{x_0}^{1} x(1-x)(x-x_0)\left\{\left[\frac{d}{dx}\varphi(x)\right]^2 - \left[\frac{d}{dx}\varphi(\mu x + 1 - \mu)\right]^2\right\} dx =$$

$$= \int_{\lambda x_0}^{x_0} x(1-x)(x_0-x)\left[\frac{d}{dx} f(x)\right]^2 dx +$$

$$+ \int_0^{x_0} x[(1-\lambda x)(x_0-\lambda x) - \tag{5.49}$$

$$-(1-x)(x_0-x)]\left[\frac{d}{dx} f(\lambda x)\right]^2 dx + \int_{x_0}^{\mu x_0+1-\mu} x(1-x)(x-x_0)\times$$

$$\times\left[\frac{d}{dx}\varphi(x)\right]^2 dx + \int_{x_0}^{1}(1-x)[(\mu x+1-\mu)(\mu x+1-$$

$$-\mu-x_0) - x(x-x_0)]\left[\frac{d}{dx}\varphi(\mu x+1-\mu)\right]^2 dx.$$

It is evident that

$$(1-\lambda x)(x_0-\lambda x) - (1-x)(x_0-x) =$$

$$= x(1-\lambda)[1+x_0-(1+\lambda)x] \geqslant 0. \tag{5.50}$$

The following inequality also holds:

$$(\mu x+1-\mu)(\mu x+1-\mu-x_0) - x(x-x_0) =$$

$$= (1-\mu)[1-\mu-x_0-(1+\mu)x^2+x(x_0+2\mu)] \geqslant 0. \tag{5.51}$$

The inequality (5.51) follows from the fact that for the function $e(x) = 1-\mu-x_0-(1+\mu)x^2+x(x_0+2\mu)$ we have the relation

$$e(1) = 0,\ e(x_0) > 0,\ e''(x) = -(1+\mu) < 0,$$

$$\max e(x) = e\left[\frac{x_0+2\mu}{2(1+\mu)}\right],\quad x_0 < x < 1.$$

We get from (5.49) by (5.50) and (5.51), that

$$J \geqslant 0. \tag{5.52}$$

Comparing (5.48) and (5.52) it can be easily seen that $u_0(x, y) \equiv 0$. Hence, in turn, there follows the uniqueness of the solution for the general mixed problem.

We shall now prove that the general mixed problem does in fact have a solution.

Denoting the right side of formulae (5.44) and (5.45) respectively by $\omega_1(x)$ and $\omega_2(x)$, these formulae can be rewritten in the form:

$$\left.\begin{aligned} &\operatorname{Re}(1-i)F'(x) = \omega_1(x) \quad 0 < x < x_0,\\ &\operatorname{Im}(1-i)F'(x) = -\omega_2(x), \quad x_0 < x < 1, \end{aligned}\right\} \tag{5.53}$$

where $F(z) = u(x, y) + iv(x, y)$ is a holomorphic function in D_1 the real part of which supplies the desired solution of the general mixed problem.

We restrict ourselves to the case when σ coincides with the semi-circle σ_0, and $x_0 > 1/2$.

In view of the fact that $u|_{\sigma_0} = 0$, function $F(z)$ can be analytically continued across σ_0 to the whole of the upper half-plane, and because of (5.53), we have

$$\left.\begin{aligned} &\operatorname{Im}(1-i)F'(x) = -\frac{1}{(2x-1)^2}\,\omega_1\left(\frac{x}{2x-1}\right),\\ &-\infty < x < 0, \quad \xi_0 = \frac{x_0}{2x_0-1} < x < \infty,\\ &\operatorname{Re}(1-i)F'(x) = \frac{x}{(2x-1)^2}\,\omega_2\left(\frac{x}{2x-1}\right), \quad 1 < x < \xi_0. \end{aligned}\right\} \tag{5.54}$$

We now introduce into our considerations the function

$$\Phi(z) = \sqrt{\left(\frac{(x_0 - z)(\xi_0 - z)}{z(1-z)}\right)}\,(1-i)F'(z),$$

which is holomorphic in the upper half-plane if by the square root we mean the single-valued branch of this function in the plane cut along the segments $[0, x_0]$, $[1, \xi_0]$ and defined to be positive in $0 < z < x_0$.

From (5.53) and (5.54) we have the boundary conditions,

$$\operatorname{Re} \Phi(x) = \begin{cases} \sqrt{\left(\frac{(x_0 - x)(\xi_0 - x)}{x(1-x)}\right)} \, \omega_1(x), \quad 0 < x < x_0, \\ -\sqrt{\left(\frac{(x_0 - x)(\xi_0 - x)}{-x(1-x)}\right)} \frac{1}{(2x-1)^2} \, \omega_1\left(\frac{x}{2x-1}\right), \\ \qquad -\infty < x < 0, \\ -\sqrt{\left(\frac{(x - x_0)(\xi_0 - x)}{x(1-x)}\right)} \omega_2(x), \; x_0 < x < 1, \\ \sqrt{\left(\frac{(x - x_0)(\xi_0 - x)}{x(x-1)}\right)} \frac{1}{(2x-1)^2} \, \omega_2\left(\frac{x}{2x-1}\right), \\ \qquad 1 < x < \xi_0, \\ \sqrt{\left(\frac{(x - x_0)(x - \xi_0)}{x(x-1)}\right)} \frac{1}{(2x-1)^2} \, \omega_1\left(\frac{x}{2x-1}\right), \\ \qquad \xi_0 < x < \infty, \end{cases}$$

for the function $\Phi(z)$, and

$$\Phi(\infty) = 0.$$

Using formula (4.56) for determining the function $\Phi(z)$, we get the expression of $F'(z)$ finally in the form

$$(1 - i) F^1(z) = \frac{1}{\pi i} \sqrt{\left(\frac{z(1-z)}{(x_0 - z)(\xi_0 - z)}\right)} \times$$

$$\times \left\{ \int_0^{x_0} \sqrt{\left(\frac{(x_0 - t)(\xi_0 - t)}{t(1-t)}\right)} \left(\frac{1}{t-z} + \frac{1-2t}{t+z-2tz}\right) \omega_1(t) \, dt - \right. \tag{5.55}$$

$$\left. - \int_{x_0}^{1} \sqrt{\left(\frac{(t - x_0)(\xi_0 - t)}{t(1-t)}\right)} \left(\frac{1}{t-z} + \frac{1-2t}{t+z-2tz}\right) \omega_2(t) dt \right\}.$$

Passing to the limit $z \to x$ in (5.55) and separating the imaginary part for $0 < x < x_0$ and the real part for $x_0 < x < 1$

we get:

$$f_x(x) + \frac{1}{\pi} \int_0^{x_0} \sqrt{\left(\frac{x(1-t)(x_0-t)(\xi_0-t)}{t(1-x)(x_0-x)(\xi_0-x)}\right)} \left(\frac{1}{t-x} + \right.$$
$$\left. + \frac{1}{t+x-2tx}\right) \times f_t(\lambda t)\, \mathrm{d}t = \varrho_1(x), \tag{5.56}$$

$$\varphi_x(x) - \frac{1}{\pi} \int_{x_0}^{1} \sqrt{\left(\frac{t(1-x)(t-x_0)(\xi_0-t)}{x(1-t)(x-x_0)(\xi_0-x)}\right)} \left(\frac{1}{t-x} - \right.$$
$$\left. - \frac{1}{t+x-2tx}\right) \times \varphi_t(\mu t + 1 - \mu)\, \mathrm{d}t = \varrho_2(x), \tag{5.57}$$

where

$$\varrho_1(x) = \psi(x) + \frac{1}{\pi} \int_{x_0}^{1} \sqrt{\left(\frac{x(1-x)(t-x_0)(\xi_0-t)}{t(1-t)(x_0-x)(\xi_0-x)}\right)} \left(\frac{1}{t-x} + \right.$$
$$\left. + \frac{1-2t}{t+x-2tx}\right) \times \varphi_t(\mu t + 1 - \mu)\, \mathrm{d}t,$$

$$\varrho_2(x) = \psi(x) - \frac{1}{\pi} \int_0^{x_0} \sqrt{\left(\frac{x(1-x)(x_0-t)(\xi_0-t)}{t(1-t)(x-x_0)(\xi_0-x)}\right)} \left(\frac{1}{t-x} + \right.$$
$$\left. + \frac{1-2t}{t+x-2tx}\right) \times f_t(\lambda t)\, \mathrm{d}t,$$

and

$$\psi(x) = \frac{1}{\pi} \sqrt{\left(\frac{x(1-x)}{(x_0-x)(\xi_0-x)}\right)} \left\{ \int_0^{x_0} \sqrt{\left(\frac{(x_0-t)(\xi_0-t)}{t(\;-t)}\right)} \times \right.$$
$$\times \left(\frac{1}{t-x} - \frac{1-2t}{t+x-2tx}\right) \psi_{2t}\left(\frac{1+\lambda}{2}\, t\right) \mathrm{d}t -$$
$$- \int_{x_0}^{1} \sqrt{\left(\frac{(t-x_0)(\xi_0-t)}{t(1-t)}\right)} \times \left(\frac{1}{t-x} + \right.$$
$$\left. \left. + \frac{1-2t}{t+x-2tx}\right) \psi_{3t}\left[\frac{t(1+\mu)+1-\mu}{2}\right] \mathrm{d}t \right\},$$
$$0 < x < x_0,$$

$$\psi(x) = \frac{1}{\pi}\sqrt{\left(\frac{x(1-x)}{(x-x_0)(\xi_0-x)}\right)}\left\{\int_0^{x_0}\sqrt{\left(\frac{(x_0-t)(\xi_0-t)}{t(1-t)}\right)}\left(\frac{1}{t-x}+\right.\right.$$

$$\left.+\frac{1-2t}{t+x-2tx}\right)\psi_{2t}\left(\frac{1+\lambda}{\lambda}t\right)\mathrm{d}t - \int_{x_0}^{1}\sqrt{\left(\frac{(t-x_0)(\xi_0-t)}{t(1-t)}\right)}\left(\frac{1}{t-x}+\right.$$

$$\left.\left.+\frac{1-2t}{t+x-2tx}\right)\psi_{3t}\left[\frac{t(1+\mu)+1-\mu}{2}\right]\mathrm{d}t\right\},$$

$$x_0 < x < 1.$$

Here we have made use of the identities

$$\frac{1}{t-x}+\frac{1-2t}{t+x-2tx} = \frac{1-t}{1-x}\left(\frac{1}{t-x}+\frac{1}{t+x-2tx}\right) =$$

$$= \frac{t}{x}\left(\frac{1}{t-x}-\frac{1}{t+x-2tx}\right).$$

In this way the proof of the existence of the solution for the general mixed problem reduces to the proof of the existence of the solutions of the integral equations (5.56) and (5.57).

By a simple change of the variables

$$\sqrt{x}\mu_1(x) = f_x(x)\sqrt{((1-x)(x_0-x)(\xi_0-x))},$$
$$\sqrt{x}h_1(x) = \varrho_1(x)\sqrt{((1-x)(x_0-x)(\xi_0-x))},$$
$$\sqrt{(1-x)}\mu_2(x) = \varphi_x(x)\sqrt{(x(x-x_0)(\xi_0-x))},$$
$$\sqrt{(1-x)}h_2(x) = \varrho_2(x)\sqrt{(x(x-x_0)(\xi_0-x))},$$

equations (5.56) and (5.57) may be rewritten in the following form:

$$\mu_1(x) + \frac{1}{\pi}\int_0^{\lambda x_0}\sqrt{\left(\frac{(\lambda-t)(\lambda x_0-t)(\lambda\xi_0-t)}{(1-t)(x_0-t)(\xi_0-t)}\right)}\left(\frac{1}{t-\lambda x}+\right.$$
$$\left.+\frac{1}{t+\lambda x-2tx}\right)\mu_1(t)\,\mathrm{d}t = h_1(x), \tag{5.58}$$

$$0 < x < x_0,$$

$$\mu_2(x) - \frac{1}{\pi}\int_{\mu x_0+1-\mu}^{1}\sqrt{\left(\frac{(t-1+\mu)(t-1+\mu-\mu x_0)(\mu\xi_0-t+1-\mu)}{t(t-x_0)(\xi_0-t)}\right)}\times$$
$$\times\left(\frac{1}{t-1+\mu-\mu x}-\right. \tag{5.59}$$
$$\left.-\frac{1}{t-1+\mu+\mu x-2x(t-1+\mu)}\right)\mu_2(t)\,\mathrm{d}t = h_2(x),$$

$$x_0 < x < 1.$$

In view of assumptions (5.39) made in connection with functions ψ_1, ψ_2 and ψ_3 we conclude that the functions $1/\sqrt{x}\,\psi(x)$, $0 < x < x_0$, and $1/\sqrt{(1-x)}\,\psi(x)$, $x_0 < x < 1$ are summable together with their squares. Therefore it is natural to look for $\mu_1(x)$ and $\mu_2(x)$ in the Hilbert space L_2.

It is easily shown that the integral operators on the left sides of (5.58) and (5.59) have norms smaller than unity for $0 < x < \lambda x_0$, $\mu x_0 + 1 - \mu < x < 1$.

This follows from

$$\frac{1}{\pi^2}\int_0^{\lambda x_0} d\xi \int_0^{\lambda x_0} \sqrt{\left(\frac{(\lambda - t)(\lambda x_0 - t)(\lambda\xi_0 - t)}{(1-t)(x_0 - t)(\xi_0 - t)}\right)}\left(\frac{1}{t - \lambda\xi} + \right.$$

$$\left. + \frac{1}{t + \lambda\xi - 2t\xi}\right)\mu_1(t)\,dt \times$$

$$\times \int_0^{\lambda x_0} \sqrt{\left(\frac{(\lambda - t_1)(\lambda x_0 - t_1)(\lambda\xi_0 - t_1)}{(1-t_1)(x_0 - t_1)(\xi_0 - t_1)}\right)}\left(\frac{1}{t_1 - \lambda\xi} + \right.$$

$$\left. + \frac{1}{t_1 + \lambda\xi - 2t_1\xi}\right)\mu_1(t_1)\,dt \leqslant$$

$$\leqslant \frac{1}{\lambda}\int_0^{\lambda x_0} \frac{(\lambda - t)(\lambda x_0 - t)(\lambda\xi_0 - t)}{(1-t)(x_0 - t)(\xi_0 - t)}\,\mu_1^2(t)\,dt \leqslant \lambda^2 \int_0^{\lambda x_0} \mu_1^2(t)\,dt.$$

The validity of the second part of our assertion can be shown in a similar way.

Therefore solutions $\mu_1(x)$, $0 < x < \lambda x_0$ and $\mu_2(x)$, $\mu x_0 + 1 - \mu < x < 1$ of equations (5.58) and (5.59) exist and are represented by singular integrals.

Substituting the expression obtained for $f_x(\lambda x)$ by inverting equation (5.56) into the right side of equation (5.57), and rearranging the left side of this last equation with respect to $\varphi_x(x)$, $\mu x_0 + 1 - \mu < x < 1$, we get a Fredholm integral equation of the second kind for $\varphi_x(x)$, $x_0 < x < 1$. Therefore the existence of the solution of the integral equation obtained in this way follows from the uniqueness of the solution for the general mixed problem.

After the existence of $\varphi_x(x)$, $x_0 < x < 1$, has been proved, the existence of $f_x(x)$, $0 < x < x_0$ follows automatically. [cf. the work of the author (9)].

The degree of smoothness of f and φ depends on the degree of smoothness of ψ_2 and ψ_3.

In the case when curve σ terminates in some small curves AA' and BB' and everywhere else is smooth and satisfies (5.47) the proof of the existence of solution for the general mixed equation reduces on the basis of the preceding paragraph by familiar methods to a Fredholm integral equation of the second kind.

The result obtained in this paragraph points to the fact that in the mixed domain D considered above the Dirichlet problem for equation (4.4) is incorrect, irrespective of the size and form of the hyperbolic part D_2 of this domain [cf. paper (9) of author].

This does not mean, naturally, that one cannot find mixed domains for which the Dirichlet problem is still possible.

For example, in the mixed domain D_3, bounded by the smooth Jordan curve σ, the characteristic AC: $x + y = 0$ and the curves σ_1 of the parabola $y = 1 - x - \sqrt{2(1-x)}$ the Dirichlet problem has a solution which is unique and continuous in D_3 with possible discontinuities occurring for the derivatives on crossing characteristic AC.

Indeed, let us look for the continuous function $u(x, y)$, in domain D_3 which, for $y \neq 0$, $y \neq x - 1$ is the solution of equation (4.4) satisfying conditions

$$u|_\sigma = \psi_1, \quad u|_{AC} = \psi_2, \quad u|_{\sigma_1} = \psi_3,$$

where ψ_1, ψ_2 and ψ_3 are given functions ψ_1 being continuous and ψ_2 and ψ_3 twice continuously differentiable, and ψ_2'' and ψ_3'' satisfy Hölder's condition.

We proved in Chapter IV that a function $u(x, y)$ given by its values on σ and on AC is uniquely determined in the closed domain $\overline{D}$, which is bounded by curves σ and characteristics AC and BC of equation (4.4) (the problem of Tricomi).

Denote by $\varphi(x)$, $1/2 \leqslant x \leqslant 1$, the boundary values of function $u(x, y)$ on the characteristic BC determined by the characteristic triangle ABC.

In the rest of the mixed domain D_3 function $u(x, y)$ is also uniquely determined from its known boundary values on the characteristic BC and on the curve σ_1 of the parabola, and the formula giving the values of this function can be written explicitly as follows:

$$u(x, y) = \varphi\left[\frac{x+y+1}{2}\right] - \varphi\left[\frac{3+\sqrt{[5-4(x-y)]}}{4}\right] +$$

$$+ \psi_3\left[\frac{1+2(x-y)+\sqrt{[5-4(x-y)]}}{4}\right].$$

§ 5. The Problem of Frankl

Suppose that we are given the equation of mixed type (equation of Chaplygin):

$$k(y)\frac{\partial^2 u}{\partial x^2} + \frac{\partial^2 u}{\partial y^2} = 0, \quad k(0) = 0,$$
$$k'(y) > 0, \quad k(-y) = -k(y) \tag{5.60}$$

and the mixed domain D, bounded by: (a) the segment $A'A$ of the y-axis $-1 \leqslant y \leqslant 1$; (b) the characteristic $A'C$ of equation

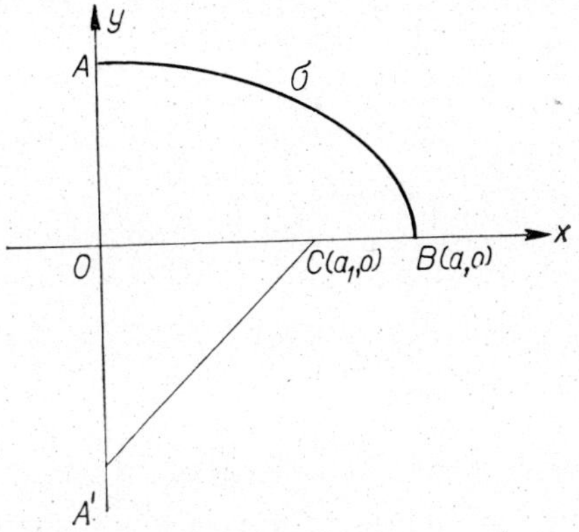

FIG. 12

(5.60) $C = C(a_1, 0)$, $a_1 > 0$; (c) segment CB of the y-axis $a_1 \leqslant x \leqslant a$ and (d) the Jordan curve σ with endpoints at points A and B, lying in the upper half-plane $y > 0$ (Fig. 12).

F. I. Frankl (5) in 1956 suggested a mixed problem which is entirely new and differs essentially from the mixed problems treated above.

This problem can be stated in the following way: find the regular solution $u(x, y)$ of equation (5.60) in domain D which is continuous in the closed domain $\overline{D}$ and satisfies the boundary conditions

$$u|_\sigma = \psi_1, \tag{5.61}$$

$$u|_{CB} = \psi_2, \tag{5.62}$$

$$\left.\frac{\partial u}{\partial x}\right|_{A'A} = 0, \tag{5.63}$$

$$u(0, y) - u(0, -y) = f(y), \; -1 \leqslant y \leqslant 1, \tag{5.64}$$

where ψ_1, ψ_2 and f are given functions.

In the present paragraph the study of Frankl's problem for the Lavrent'ev equation (4.4) is given.

In this case $A'C$ coincides with a segment of the characteristic $x - y = 1$ of equation (4.4). We shall put an additional restriction on the curve σ, apart from the usual smoothness condition (Liapunov condition), namely we shall demand the inequality

$$\frac{dy}{ds} \geqslant 0, \tag{5.65}$$

where $x = x(s)$, $y = y(s)$ are parametric equations and s is the curve length of σ measured from $B(a, 0)$. Without loss of generality it can be assumed that $\psi_1 = \psi_2 = 0$. We shall demand the differentiability of $f(y)$ up to the second order and that its second derivative should satisfy the Hölder condition.

Under these assumptions there exists a unique function $u(x, y)$ with the following properties: (1) $u(x, y)$ is the solution of equation (4.4) for $y \neq 0$, and for $y \neq -x$; (2) it is continuous in the closed domain $\overline{D}$; (3) the partial derivatives $\partial u/\partial x$ and $\partial u/\partial y$ are continuous in the closed domain $\overline{D}$ except perhaps on segment OC' of the characteristic $x + y = 0$, $O = O(0, 0)$,

$C' = C'(^1/_2, -^1/_2)$ and at points A, A', C, B, where they may become infinite of order lower than unity; (4) $u(x, y)$ satisfies the boundary conditions (5.61)—(5.64).

The general solution of equation (4.4) satisfying the conditions

$$\frac{\partial u}{\partial x}\bigg|_{OA'} = 0, \quad u|_{OA'} = \varphi(-y) + f(y),$$

in triangle $A'OC'$ has the following form

$$u(x, y) = \frac{1}{2}\varphi(-x-y) + \frac{1}{2}\varphi(x-y) + \\ + \frac{1}{2}f(x+y) + \frac{1}{2}f(y-x), \tag{5.66}$$

where $\varphi(y)$ means function $u(0, y)$, $0 \leqslant y \leqslant 1$.

From formula (5.66) we have

$$u(x, -x) = \frac{1}{2}\varphi(0) + \frac{1}{2}\varphi(2x) + \\ + \frac{1}{2}f(-2x), \quad 0 \leqslant x \leqslant \frac{1}{2}. \tag{5.67}$$

The general solution of equation (4.4) satisfying the condition (5.67) is given by the formula

$$u(x, y) = \Phi(x+y) + \frac{1}{2}\varphi(x-y) + \frac{1}{2}f(y-x) \tag{5.68}$$

in the triangle $OC'C$ where $\Phi(t)$ is an arbitrary twice continuously differentiable function in $0 < t < 1$, which is continuous in $0 \leqslant t \leqslant 1$, and $\Phi(0) = 1/2\,\varphi(0)$.

From (5.68) we have

$$u(x, 0) = \Phi(x) + \frac{1}{2}\varphi(x) + \frac{1}{2}f(-x), \quad 0 \leqslant x \leqslant 1. \tag{5.69}$$

By $v(x, y)$ we denote the harmonic conjugate of $u(x, y)$ in the elliptic part D_1 of domain D and let $v(0, 0) = 0$.

We conclude from (5.63) that $v\big|_{OA} = 0$. Since by hypothesis $\partial u/\partial y$ is continuous across OC, we find from (5.68) and the equation $\partial v/\partial x = -\partial u/dy$ that

$$v(x, 0) = -\Phi(x) + \frac{1}{2}\varphi(x) + \frac{1}{2}f(-x) \quad 0 \leqslant x \leqslant 1. \tag{5.70}$$

It is now easy to show that the homogeneous problem $(f = 0)$ has only the trivial solution.

Indeed, let $u_0(x, y)$ be the solution of the homogeneous problem, and $v_0(x, y)$ its harmonic conjugate such that $v_0(0, 0) = 0$. The function $F_0(z) = u_0(x, y) + iv_0(x, y)$ is holomorphic in the domain D_1 and is continuous in the closed domain $\overline{D}_1$. Therefore

$$-\int_{\sigma} v_0^2 \, dz - i \int_0^1 u_0^2(0, y) \, dy + \int_0^1 (u_0 + iv_0)^2 \, dx - \int_1^a v_0^2 \, dx = 0. \tag{5.71}$$

Equating the real part of the left side of (5.71) to zero and using (6.69) and (5.70) we obtain the relation

$$-\int_{\sigma} v_0^2 \frac{dy}{ds} \, ds - \int_0^1 \varphi_0^2(y) \, dy - 2\int_0^1 \left(\Phi^2 - \frac{1}{4} \varphi_0^2\right) dx = 0,$$

$$\varphi_0(y) = u_0(0, y).$$

Hence, according to (5.65) $\Phi(x) = \varphi_0(x) = 0$ and also $u_0(x, y) = 0$ everywhere in the mixed domain D.

We proceed now to prove the existence of the solution.

We note that because of (5.69) and (5.70), we have

$$u(x, 0) + v(x, 0) = \gamma(x), \quad 0 \leqslant x \leqslant 1, \tag{5.72}$$

where

$$\gamma(x) = \varphi(x) + f(-x). \tag{5.73}$$

The function $F(z) = u(x, y) + iv(x, y)$ which is holomorphic inside D_1 and continuous in $\overline{D}_1$ is uniquely determined by the boundary conditions:

$$u|_{\sigma} = 0, \quad u(x, 0) = 0, \quad 1 \leqslant x \leqslant a, \quad v(0, y) = 0, \quad 0 \leqslant y \leqslant 1,$$
$$u(x, 0) + v(x, 0) = \gamma(x), \quad 0 \leqslant x \leqslant 1.$$

In particular, when $a = 1$, σ coincides with the curve of the circle $x^2 + y^2 = 1$, $y \geqslant 0$, $x \geqslant 0$, and the function $F(z)$ can be easily expressed as follows:

$$F(z) = \frac{1 + i}{\pi i} \int_0^1 z \sqrt{\left(\frac{t(1 - z^2)}{z(1 - t^2)}\right)} \left(\frac{1}{t^2 - z^2} - \frac{1}{1 - t^2 z^2}\right) \gamma(t) \, dt. \tag{5.74}$$

Substituting iy for z in formula (5.74) and taking (5.64) and (5.73) into account we get for the determination of the unknown function $\varphi(y)$ the integral equation:

$$\varphi(y) - \frac{\sqrt{2}}{\pi}\int_0^1 y\sqrt{\left(\frac{t(1+y^2)}{y(1-t^2)}\right)}\left(\frac{1}{t^2+y^2} - \frac{1}{1+t^2y^2}\right)\varphi(t)\,dt = \psi(y), \tag{5.75}$$

where

$$\psi(y) = \frac{\sqrt{2}}{\pi}\int_0^1 y\sqrt{\left(\frac{t(1+y^2)}{y(1-t^2)}\right)}\left(\frac{1}{t^2+y^2} - \frac{1}{1+t^2y^2}\right)f(-t)\,dt \tag{5.76}$$

is a known function.

We introduce the notation $\sqrt{y}\,\varphi(y) = \mu(y)$.

For determining the unknown $\mu(y)$ we get according to (5.75) the integral equation:

$$\mu(y) - \frac{\sqrt{2}}{\pi}\int_0^1 y\sqrt{\left(\frac{1+y^2}{1-t^2}\right)}\left(\frac{1}{t^2+y^2} - \frac{1}{1+t^2y^2}\right)\mu(t)\,dt = \sqrt{y}\,\psi(y) \tag{5.77}$$

with positive kernel such that

$$\frac{\sqrt{2}}{\pi}\int_0^1 y\sqrt{\left(\frac{1+y^2}{1-t^2}\right)}\left(\frac{1}{t^2+y^2} - \frac{1}{1+t^2y^2}\right)dt = \frac{1-y}{\sqrt{2}}. \tag{5.78}$$

From (5.78) it is evident that the integral equation (5.77) has a solution which can be constructed by ordinary means. Having found the function φ, the solution of Frankl's problem in the considered case can be obtained by integration.

We return now to the general case. Owing to the fact that $v\,|_{OA} = 0$ the function $F(z)$ can be extended analytically into the domain D_1^*, which is the reflection in the imaginary axis of the domain D_1. The union of the open segment OA and the domains D_1 and D_1^* we shall now denote by D^*. The required function $F(z)$ is holomorphic in domain D^*, continuous in the closed domain $\overline{D}^*$ and satisfies the boundary conditions

$$u|_{\sigma} = 0, \; u|_{\sigma^*} = 0,$$

$$\left.\begin{aligned} u(x, 0) = 0, \; -a \leqslant x \leqslant -1, \; u(x, 0) - \\ - v(x, 0) = \gamma(-x), \\ -1 \leqslant x \leqslant 0, \; u(x, 0) + v(x, 0) = \\ = \gamma(x), \; 0 \leqslant x \leqslant 1, \\ u(x, 0) = 0, \; 1 \leqslant x \leqslant a. \end{aligned}\right\} \quad (5.79)$$

Here σ^* denotes the curve which is the reflexion of σ in the imaginary axis.

Since domain D^* is symmetric with respect to the imaginary axis and contains the segment OA, in the conformal mapping of this domain onto the semi-circle k_1: $\xi^2 + \eta^2 < 1, \; \eta > 0$, of the plane $\zeta = \xi + i\eta$, which takes points $A(0, 1)$, $C^*(-1, 0)$, $C(1, 0)$ into the points $A_1(0, 1)$, $C_1(-1, 0)$, $C_1(1, 0)$, the symmetry with respect to the imaginary axis is preserved.

We denote by $\Phi(\zeta)$ the function $F[\omega^{-1}(\zeta)]$, where $\zeta = \omega(z)$ is the function realizing the conformal mapping.

For determining the function $\Phi(\zeta)$, which is holomorphic in the semi-circle k_1 and continuous in k_1 we have from (5.79) the conditions

$$\operatorname{Re} \Phi|_L = 0, \quad \operatorname{Re}(1 + i)\Phi|_{C_1^* O_1} = \gamma[-\omega^{-1}(\xi)], \qquad (5.80)$$

$$\operatorname{Im}(1 + i)\Phi|_{O_1 C_1} = \gamma[\omega^{-1}(\xi)],$$

where L is the semi-circle $|\zeta| = 1, \; \eta \geqslant 0, \; O_1 = O_1(0, 0)$.

In the ζ-plane the function $\Phi(\zeta)$ is again determined by formula (5.74). From this formula an integral equation is obtained for the determination of the function φ, which is equivalent to the Frankl problem and for which the existence of solution is made certain on the basis of the uniqueness of solution for this problem. [cf. A. V. Bitsadze (10)].

§ 6. Short Indication of some Important Generalizations and Applications

(1) In the preceding paragraphs of this chapter we limited ourselves to the consideration of the Lavrent'ev equation (4.4) only.

The method of solution of mixed problems applied in these paragraphs, may also be used in the case of equations more general than (4.4). For instance, the method indicated in § 5 for establishing the uniqueness of solution of the Frankl problem can also be successfully applied also in the case of equation (5.60) [cf. A. V. Bitsadze (11), J. V. Devengtalj (1)].

Indeed, let us denote by $v\,(x,\ y)$ the function which satisfies together with $u\,(x,\ y)$, the solution of the Frankl problem for the system of equations:

$$k\,(y)\,\frac{\partial u}{\partial x} - \frac{\partial v}{\partial y} = 0, \qquad \frac{\partial u}{\partial y} + \frac{\partial v}{\partial x} = 0. \tag{5.81}$$

Under the condition $v\,(0,\ 0) = 0$, function $v\,(x,\ y)$ is uniquely determined from the relation (5.81) by the aid of $u\,(x,\ y)$. In addition we shall demand that the boundary of D should satisfy conditions sufficient for the continuity of $v\,(x,\ y)$ in the closed domain $\overline{D}$.

Let $a = a_1$. For every closed rectifiable curve C, contained in D, we have from (5.81)

$$\int_C (ku^2 - v^2)\,\mathrm{d}y + 2uv\,\mathrm{d}x = 0. \tag{5.82}$$

Equation (5.82) is valid whenever C coincides with the complete boundary of domain D. By means of this identity we show that the homogeneous Frankl problem ($\psi_1 = \psi_2 = f = 0$) has only the trivial solution.

Indeed, since by (5.63) $v\,(0,\ y) = 0$, $-1 \leqslant y \leqslant 1$, we may use (5.60), (5.61) and (5.64) to write

$$\int_\sigma v^2\,\frac{\mathrm{d}y}{\mathrm{d}s}\,\mathrm{d}s + \int_{A'B} [\sqrt{(-ku)} - v]^2\,\mathrm{d}y = 0. \tag{5.83}$$

From (5.65) we see that every term on the left side of (5.83) vanishes. Therefore on those parts of the curve σ, where $dy/ds > 0$ we have $u = v = 0$. Hence, due to the ellipticity of the system (5.81) in the elliptic part D_1 of the mixed domain D it follows that $u = v = 0$ in domain D_1 [Cerleman (1), Vekua (2)].

$$u(x, 0) = 0, \left.\frac{\partial u(x, y)}{\partial y}\right|_{y=0} = 0, \quad 0 \leqslant x \leqslant a. \tag{5.84}$$

It presents no difficulty to show that the singular Cauchy problem (5.84) for system (5.81) cannot have a non-zero solution under sufficiently general conditions concerning $k((y)$, in the hyperbolic part of domain D. In this way $u(x, y) = 0$ everywhere in D.

(2) During the study of problem M of the general mixed problem and of Frankl's problem we placed quite strong restrictions on the elliptic part σ of the boundary of the mixed domain [cf. conditions (5.3), (5.47) and (5.65)]. With the application of the method suggested by K. Moravecz in proving the uniqueness of the solution for certain mixed problems (1, 2) these conditions can be weakened.

It would be most desirable to find ways of proving the existence and the uniqueness of the solution for the said problems without placing any restriction on curve σ apart from the usual smoothness requirements. In this context it would be interesting to clear up the question whether there is an extremal principle also for these problems analogous to the one proved in Chapter 4.

(3) When in the introduction we spoke about the importance of equations of mixed type in applications we had in mind first of all the application of the theory of these equations to the flow of compressible fluids in hydrodynamics and to questions arising in the theory of shells.

S. A. Chaplygin has pointed out that the building up of the theory of gas flow is closely connected with the study of the equation

$$k(y)\frac{\partial^2 u}{\partial x^2} + \frac{\partial^2 u}{\partial y^2} = 0, \tag{5.85}$$

which, at present, is named the Chaplygin equation (1). Coefficient k is the known function of y, it is supposed to be positive for $y < 0$, negative for $y < 0$. The case $k(y) < 0$ corresponds to subsonic and the case $k(y) > 0$ to supersonic gas flows. In the way of additional hypothesis it can be assumed that $k(y) = y$ [cf. for example, Tricomi (7)] or $k(y) = \operatorname{sgn} y$ (L. I. Sedov (1)].

For studying gas flow having velocity approximately equal to the velocity of sound we have to deal with equation (4.1) in a mixed domain.

In his first investigations (1—4) dealing with equation (4.1) Tricomi, apparently, started from a pure mathematical interest.

F. I. Frankl first drew attention to the fact that the Tricomi problem and several other mixed problems, dealt with above are closely related with the study of gas flow with nearly sonic speeds, (1—6).

I. N. Vekua (2) noticed that equations of mixed type are encountered also in the theory of shells in the case when the principal curvature of one shell changes sign.

A sufficiently complete survey of the mathematical formulation of mixed problems encountered in the study of supersonic gas flow can be found in the monograph by Bers (2) published a short while ago.

The replacement of the Chaplygin equation by the equations of Tricomi or Lavrent'ev may be and is crude from the point of view of actual gas-dynamic applications; nevertheless to us it appears natural that the first mathematical investigations aiming at the problem of mixed equations dealt precisely with these equations as models.

(4) The investigation of mixed problem for the Chaplygin equation or for more general equations (and systems) of mixed type is of great theoretical and practical importance.

Certain results in this direction for the Chaplygin equation were announced by K. I. Babenko. The paper of Moravetz (4) must also be mentioned in which a new approach to the proof of existence and uniqueness of solution in case of the mixed problem for the Chaplygin equation was suggested. We shall give below a short account of this work.

Suppose we are given the non-homogeneous Chaplygin equation

$$k(y)\frac{\partial^2 \psi}{\partial x^2} + \frac{\partial^2 \psi}{\partial y^2} = g(x, y), \tag{5.86}$$

where $k(y)$ is a differentiable function satisfying conditions $k(0) = 0$, $k'(y) > 0$ for both positive and negative values of y. As to the restriction placed on function $g(x, y)$ we shall discuss them later.

Denote by D the finite singly-connected mixed domain with boundary $B = C_1 + C_0 + C_2 + \gamma_1 + \gamma_2$ (Fig. 13). The curve C_0 lies in the upper semi-plane cutting the axis $y = 0$ in points $X_1(x_1, 0)$, $X_2(x_2, 0)$, $x_1 < 0 < x_2$ and has a piece-wise con-

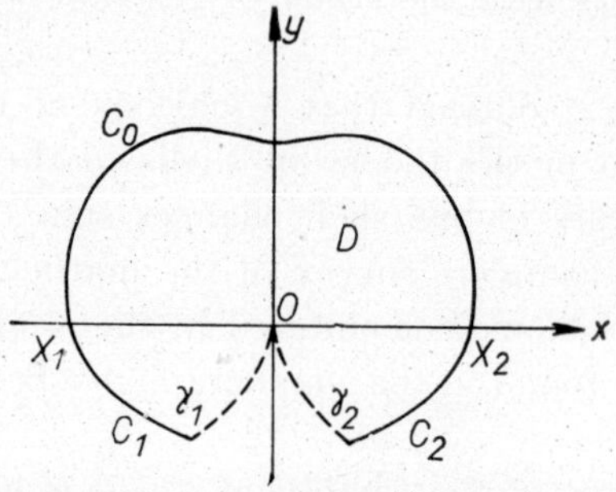

FIG. 13

tinuous tangent. Curves C_1 and C_2 lie in the lower semi-plane. They start from points X_1 and X_2 respectively and intersect the characteristics γ_1 and γ_2 of equation (5.86) starting from the origin of the coordinate system; on C_1 and C_2 we have the inequality:

$$k(y)\left(\frac{dy}{dx}\right)^2 + 1 \geqslant 0. \tag{5.87}$$

The problem consists of finding the solution $\psi(x, y)$ of equation (5.86) in the domain D, which satisfies on $C = C_1 + C_0 + C_2$ the condition:

$$\psi = 0. \tag{5.88}$$

This problem was originally stated by F. I. Frankl. (2).

By obvious alterations in the considerations contained in § 4 of this chapter, this problem, for $k(y) = y$, can be investigated by means of the method of integral equations.

Introducing the notation $\partial\psi/\partial x = u_1$, $\partial\psi/\partial y = u_2$, the above problem of Morawetz leads to the determination of vector $u = (u_1, u_2)$, in domain D satisfying the conditions:

$$Lu = g \quad \text{in } D,$$

$$u_1 \frac{dx}{ds} + u_2 \frac{dy}{ds} = 0 \text{ on } C. \tag{5.89}$$

Here the notation

$$(Lu)_1 = k(y) \frac{\partial u_1}{\partial x} + \frac{\partial u_2}{\partial y}, \quad (Lu)_2 = \frac{\partial u_1}{\partial y} - \frac{\partial u_2}{\partial x}, \quad g = (g, 0)$$

is used.

Let H_* be the Hilbert space of the measurable vectors $u = (u_1, u_2)$ in domain D having finite norm.

$$\|u\|_* = \left(\iint_D (ru_1^2 + u_2^2)\, dx\, dy\right)^{1/2},$$

where $r^2 = x^2 + y^2$.

The scalar product of two vectors $u \in H_*$, $v = (v_1, v_2) \in H_*$ is given by the formula:

$$(u, v)_* = \iint_D (ru_1 v_1 + u_2 v_2)\, dx\, dy.$$

With this definition of the norm, the point $O(0, 0)$ (from which the boundary characteristics γ_1, γ_2 commence) which lies on the curve of parabolic degeneracy $y = 0$ possesses a special position. This circumstance is partly the counterpart of the fact that the y-derivative of the solution of Tricomi's problem may become infinite at the endpoint $B(1, 0)$ of the characteristic CB which is free of boundary conditions.

Denote by H the set of continuously differentiable vectors

$w = (w_1, w_2)$, satisfying the conditions

$$\begin{aligned} & w = (0, 0) \text{ for } r = 0; \\ & w_1 = 0 \quad \text{on } C; \\ & \sqrt{-k}\, w_1 + w_2 = 0 \text{ on } \gamma_1, \\ & \sqrt{-k}\, w_1 + w_2 = 0 \text{ on } \gamma_2; \end{aligned} \tag{5.90}$$

$$\iint\limits_D \left(\frac{1}{r} (Lw)_1^2 + (Lw)_2^2\right) dx\, dy < \infty. \tag{5.91}$$

Morawetz calls vector $u \in H_*$ the weak solution of problem (5.89) if for every $w \in H$ it is true that

$$(w, g) = -(Lw, u), \tag{5.92}$$

where

$$(w, g) = \iint\limits_D (w_1 g_1 + w_2 g_2)\, dx\, dy.$$

If a weak solution $u(x, y)$ is continuously differentiable it is a solution also in the ordinary sense.

Indeed, applying Green's familiar formula, we get

$$-(Lw, u) = (Lu, w) - \int\limits_B w_1 (ku_1\, dy - u_2\, dx) + \\ + w_2 (u_1\, dx + u_2\, dy). \tag{5.93}$$

On the characteristic γ_1 we have $dx = \sqrt{-k}\, dy$. Therefore, the integral on the right side of (5.93) assumes on γ_1 the form

$$\int\limits_{\gamma_1} \left[\sqrt{(-k)}\, u_1 + u_2\right] \left[\sqrt{(-k)}\, w_1 + w_2\right] dy.$$

This expression is equal to zero due to (5.90). In a similar fashion we can verify that the integral over γ_2 on the right side of (5.93) also vanishes.

In this way, for every $w \in H$ equation (5.93) becomes

$$(w, g) = (Lu, w) - \int\limits_C w_2 (u_1\, dx + u_2\, dy).$$

Hence it follows directly that the vector u is indeed a solution of problem (5.89).

With the purpose of proving the existence of a weak solution the new Hilbert space H^* of the measurable vectors $t = (t_1, t_2)$ with the finite norm

$$\|t\|^* = \left(\iint_D \left(\frac{1}{r} t_1^2 + t_2^2\right) \mathrm{d}x \, \mathrm{d}y\right)^{1/2}$$

is introduced.

In space H^* the scalar product of vectors t and $s = (s\ ,\ s_2)$ is given by the formula:

$$(t,\ s)^* = \iint_D \left(\frac{1}{r} t_1 s_1 + t_2 s_2\right) \mathrm{d}x \, \mathrm{d}y.$$

From (5.91) we conclude that $Lw \in H^*$.

We accept now without proof that for every $w \in H$ we have

$$\|w\|_* \leqslant B \|Lw\|^*, \tag{5.94}$$

where B is a fixed constant (independent of w).

By the Schwartz inequality we have:

$$|(w, f)| \leqslant \|w\|_* \cdot \|g\|^*.$$

Hence, from (5.94), we get

$$|(w, g)| \leqslant B \|Lw\|^* \|g\|^*. \tag{5.95}$$

For a fixed vector $g \in H^*$ the expression (w, g) is a (bounded) linear functional on Lw:

$$(w, g) = G(Lw).$$

The functional $G(s)$ is defined only for elements of the form $s = Lw \in H^*$, where as yet $w \in H$. But its definition can be extended to all points in space H^*.

By (5.95) we may write

$$|G(Lw)| \leqslant B_1 \|Lw\|^*,$$

where $B_1 = B_1(g)$.

From functional analysis we have the following theorem: if $G(s)$ is a (bounded) linear functional defined over the whole of space H^*, then an element $t \in H^*$, can be found such that

$$G(s) = (s, t)^*$$

or what is the same thing

$$(w, g) = G(Lw) = (Lw, t)^*,$$

where w is any element of H.

Consider now vector u with components $u_1 = -t_1/r$, $u_2 = -t_2$, where $t = (t_1, t_2) \in H^*$.

Owing to the equation

$$\iint_D (ru_1^2 + u_2^2)\,\mathrm{d}x\,\mathrm{d}y = \iint_D \left(\frac{t_1^2}{r} + t_2^2\right)\mathrm{d}x\,\mathrm{d}y$$

we conclude that $u \in H_*$.

Therefore we have

$$(w, g) = (Lw, t)^* = \iint_D \left(\frac{1}{r}(Lw)_1 t_1 + (Lw)_2 t_2\right)\mathrm{d}x\,\mathrm{d}y =$$

$$= -\iint_D ((Lw)_1 u_1 + (Lw)_2 u_2)\,\mathrm{d}x\,\mathrm{d}y = -(Lw, u),$$

from which it follows directly that vector $u \in H_*$ is the weak solution we were looking for.

It follows from (5.94) that the adjoint equation $w \in H$, $Lw = 0$ also has a unique solution.

The proof of the extremely important inequality (5.94) will not be reproduced here.

In the proof of this inequality Morawetz needed to make further restrictions concerning the boundary of domain D in the form of the following inequalities:

$$x\,\mathrm{d}y > k_0\,\mathrm{d}s \text{ along } C_1 \text{ and } C_2,$$

$$y^{-1/2}(x\,\mathrm{d}\mu - \mu\,\mathrm{d}x) \geqslant k_0\,\mathrm{d}s > 0 \text{ along } C_0.$$

Here k_0 is some positive constant and

$$\mu = \int_0^y \sqrt{k}\,\mathrm{d}y.$$

Unfortunately these are quite strong restrictions and it would be desirable to get rid of them. At the same time it is important to identify the conditions which are sufficient for a solution which is weak in the sense of Morawetz to become a solution in the ordinary sense.

Beside the works of the authors mentioned in the text of this book, a whole series of publications has appeared during recent years (Agmon (1), N. I. Bakievich (1), P. G. Barancev (1—2), Van Guan-in (1), Weinberger (1), Germain and Bader (4), Germain (2, 3) I. L. Karol (2—4), K. Morawetz (3)] concerning which information can be found in the Bibliography following.

The methods which are at our disposal at present for investigating equations of mixed type are more or less special in their character. This is partly responsible for the difficulties which arose even in connection with the statement of mixed problems for equations of mixed type in multi-dimensional domains.

REFERENCES

AGMON, S.
(1) (1955) Boundary Value Problems for Equations of the Mixed Type. *Convergo internaz.*, Equazioni lineari alle derivate parziali, Roma, 54—68.

AGMON, S., NIRENBERG, L., and PROTTER, M. H.
(1) (1955) A Maximum Principle for a Class of Hyperbolic Equations and Applications to Equations of Mixed Elliptic-hyperbolic Type. *Commun. Pure Appl. Mat.*, *6*, 455—470.

BABENKO, K. I.
(1) (1952) On the Theory of Equations of the Mixed Type. *Doctoral thesis* (Library of the Mat. Inst. Akad. Nauk. SSSR),
(2) (1953) On the Theory of Equations of the Mixed Type. *Usp. Mat. Nauk, VIII*, 2 (54), 160.

BAKIEVICH, N. I.
(1) (1957) Some Boundary Value Problems for Equations of the Mixed Type in a Strip and in the Half-plane. *Dokl. Akad. Nauk SSSR 112*, 793—796.

BARANTSEV, P. G.
(1) (1957) A Mixed Problem for the Equation $\psi_{\sigma\sigma} - k(\sigma)\,\psi_{\theta\theta} = 0$ with Values given on the Curve $\theta = \theta(s)$ *Dokl. Akad. Nauk SSSR 114*, 919—922,
(2) (1957) The Boundary Value Problem for the Equation $\psi_{\sigma\sigma} - k(\sigma)\,\psi_{\theta\theta} = 0$ with Values given on the Characteristics and on the Straight Lines σ Const. *Dokl. Akad. Nauk SSSR, 113*, 955—958,

BERGMAN, S.
(1) (1952) On Solutions of Linear Partial Differential Equations of Mixed Type. *Amer. J. Math.*, *74*, 2, 444—474.

BEREZIN, I. S.
(1) (1949) On the Cauchy Problem for Linear Second Order Equations with Values given on the Lines of Parabolicity. *Mat. Symposium 24*, 301—320.

BERS, L.
(1) (1950) On the Continuation of a Potential Gas Flow across the Sonic Line. *Nat. Adv. Com. Aeronautics*, Tech. Not., 2058.
(2) (1958) Mathematical Aspects of Subsonic and Transonic Gas Dynamics. *Surv. Appl. Math.* N. Y., 4.

BITSADZE, A. V.
(1) (1948) On the Uniqueness of the Solution to the Dirichlet Problem. *Usp. Mat. Nauk, III*, 6 (28) 211.
(2) (1957) On Systems of Second Order Elliptic Partial Differential Equations. *Dokl. Akad. Nauk SSSR, 112*, 6, 983—986.
(3) (1944) Boundary Value Problems for Elliptic Systems of Linear Differential Equations. *Soobshch. Akad. Nauk. SSSR. 8*, 761—770.

(4) (1950) On Some Problems of Mixed Type. *Dokl. Akad. Nauk. SSSR 70*, 4, 561—564.

(5) (1951) On the Problem of Equations of the Mixed Type. *Doctoral thesis*, (Library of the Mat. Inst. Akad. Nauk. SSSR).

(6) (1953) On the Problem of Equations of the Mixed Type. *Trud. Mat. Inst. Akad. Nauk SSSR, XLI.*

(7) (1957) On an Elementary Method for the Solution of Certain Boundary Value Problems in the Theory of Holomorphic Functions and of Related Singular Integral Equations. Usp. mat. nauk *XII.* 5 (77), 185-190.

(8) (1956) On the Problem of Equations of the Mixed Type in Multidimensional domains. Dokl. Akad. Nauk SSSR 110, 6, 901-902.

(9) (1953) The Incorrectness of Dirichlet Problem for Equations of Mixed Typa. *Dokl. Akad. Nauk SSSR* 122, 2. 167-170,

(10) (1956) On a problem of Frankl, *Dokl. Akad. Nauk SSSR* 108, 6, 1091—1094.

(11) (1957) On the Uniqueness of the Frankl Problem for Chaplygin Equations *Dokl. Akad. Nauk SSSR 119*, 3, 375—376.

CARLEMAN, T.

(1) (1939) Sur un Problème d'Unicité pour les systèmes d'Equations aux Derivées Partielles à Deux Variables Indépendantes. *Arkiv. f. M. A. O. F. 26B*, 17, 1—9.

CHAPLYGIN, S. A.

(1) (1933) On Jets of Gases. Complete Collected works, *II.*

CHI MIN'YU

(1) (1958) On the Cauchy Problem for a Class of Hyperbolic Equations with Initial Data on the Line of Parabolic Singularity. *Acta Mat. Sinica, 8*, 4, 521—530.

CIBRARIO, M.

(1) (1932) Sulla Reduzione a Forma Canonica Delle Equazioni Lineari Alle Derivate Parziali di Secondo Ordine di tipo Misto Iperbolo- paraboliche. *Rend. Cir. Mat. Palermo, LVI,*

(2) (1932) Sulla Reduzione a Forma Canonica Delle Equazioni Lineari Alle Derivate Parziali di Secondo Ordine di Tipo Misto. *Rends. Lombardo, 65*, 889—906.

(3) (1932—1933) Alcuni Teoremi di Esistenza e di Unicita per l'Equazione $xz_{xx} + Z_{yy} = 0$ *Atti R. Acc. Torino, 68.*

CONTI, R.

(1) (1950) Sul Problema di Cauchy per l'Equazione $q^{2x} k^2 (x, y)$ $\partial^2 z/\partial x^2 - \partial^2 z/\partial y^2 = f(x, y, z, z_x, z_y)$. Con i Data Sulla Linea Parabolica. *Ann. Math., 31*, 303—326.

DEVENGTAL', YU. V.

(1) (1958) On the Existence of Solution for a Problem of F. I. Frenkel. *Dokl. Akad. Nauk. SSSR 119*, 1, 15—18.

DIN SYA-SI

(1) (1955) Differential Equations of the Mixed Type. *Acta Mat. Sinica*, 193—204.

FILIPPOV, A. G.

(1) (1951) On the Application of Finite Differences to the Solution of the Problem of Tricomi. *Izv. Akad. Nauk SSSR ser. mat. 21*, 13—88.

FRANKL', F. I.

(1) (1945) Cauchy's Problem for Partial Differential Equations of Mixed Elliptico-Hyperbolic Type with Initial Data on the Parabolic Line *Izv. Akad. Nauk SSSR, ser. mat. 8*, 5, 195—224.

(2) (1945) Problems of Chaplygin for Mixed Sub- and Supersonic Flows. *Izv. Akad. Nauk SSSR ser. mat. 9*, 2, 121—143.

(3) (1947) On a Family of Partial Solutions to the Darboux-Tricomi Equation and its Application to the Approximate Calculation of the Critical Current in a Given Plane-Parallel Laval-nozzle. *Dokl. Akad. Nauk, SSSR 56*, 683—686.

(4) On a New Boundary Value Problem for the Equation $yz_{xx} + z_{yy} = 0$. *Uch. zap. Mosk. Gos. Univ. mekhanika, III*, 99—116.

(5) (1956) Subsonic Flow about a Profile with a Supersonic Zone. *Zh. prikl. mat. i mekh. XX*, 2, 196—202.

(6) (1951) Two Gas-dynamical Applications of the boundary Value Problem of Lavrent'ev—Bitsadze. *Vest. Mosk. Gos. Univ. 11* 3—7.

FRIEDRICHS, K. O.

(1) (1958) Symmetric Positive Linear Differential Equations. *Commun. Pure Appl. Math. XI*, 3, 333—418.

GELLERSTEDT, S.

(1) (1935) Sur un Problème aux Limites pour une Equation Linéaire aux Dérivées Partielles du Second Ordre de Type Mixte. Thèse, Uppsala,

(2) (1936) Sur un Problème aux Limites pour l'Equation $y^{2s}z_{xx} + z_{yy} = 0$ *Arkiv, f. M. A. O. F. 25A*, 10.

(3) (1936) Quelques Problèmes Mixtes pour l'Equation $y^m z_{xx} + z_{yy} = 0$ *Arkiv, f. M. A. O. F., 26A*, 3.

GERMAIN, P.

(1) (1950) Nouvelles Solutions de l'Equation de Tricomi *C. R. Acad. Sci.* Paris, *231*, 1116—1118.

(2) (1954) Remarques sur les Propriétes Qualitatives des caractéristiques des equations aux Déricées Partielles du Type Mixte, *Mém. Acad. Roy. Belg. cl. sci.*, *28*, 28—36.

(3) (1955) Remarks on Transforms and Boundary Value Problems. *J. Rat. Mech. a anal.*, *4*, 925—941.

GERMAIN, P. and BADER, R.

(1) (1951) Sur le Problème de Tricomi. *C. R. Acad. Sci.*, Paris, *232* 365—463.

(2) (1952) Recherches sur un Equation du Type Mixte. Problèmes Elliptiques et Hyperboliques Singuliers pour une Equation du Type Mixte. Note technique O. N. E. R. A.,

(3) (1952) Sue Quelques Problémes Relatifs à l'Equation du Type Mixte de Tricomi.

(4) (1953) Solutions Élémentaires de Certaines Équations aux Dérivées Partielles du Type Mixte. *Bull. Soc. Math. France, 81*, 145—174.

HAACK, W. and HELLWIG, G.

(1) (1954) Lineare Partielle Differentialgleichung zweiter Ordnung von gemischtem Typus. *Arch. Math. 5*, 60—76.

HAUSDORFF, F.

(1) (1932) Zur Theorie der linearen metrischen Räume. *J. F. Reine und angew. Math. 167*, 294—311. Translated into Russian (cf. Teoriya mnozhestv, dobavleniye), 1937. United Scientific and Technical Press.

HELLWIG, G.

(1) (1954) Über Partielle Differentiagleichungen zweiter Ordnung von gemischtem Typus. *Math. Zeitschrift, 61*, 24—46.

(2) (1956) Anfangswertprobleme bei partiellen Differentialgleichun-

gen mit Singularitäten. *J. Rat. Mech. a Anal.*, *5*, 2, 395—418.

Kapilevich, M. B.

(1) (1952) On an Equation of Mixed Elliptic-hyperbolic Type. *Mat. Simposium*, *30*, 1, 11—38.

Karapetyan, K. I.

(1) On the Cauchy Problem for a Hyperbolic Equation which is Singular on the Initial Plane. *Dokl. Akad. Nauk SSSR 106*, 6, 963—966.

Karmonov, V. G.

(1) (1954) On a Boundary Value Problem for an Equation of the Mixed Type. *Dokl. Akad. Nauk SSSR 95*, 3, 439—442.

(2) (1958) On the Existence of Solutions of Certain Boundary Value Problems for an Equation of the Mixed Type. *Izv. Akad. Nauk SSSR ser. mat. 22*, 1, 117—134.

Karol', I. L.

(1) (1955) On the Theory of Boundary Value Problems for an Equation of the Mixed Elliptic-hyperbolic Type. *Mat. Symposium*, *38* (80), 3, 261—282.

(2) (1953) On a Boundary Value Problem for a Mixed Equation of the Elliptic-hyperbolic Type. *Dokl. Akad. Nauk SSSR 88*, 197—200.

(3) (1953) On the Theory of Equations of the Mixed Type. *Dokl. Akad. Nauk, SSSR 88*, 397—400.

(4) (1955) Boundary Value Problems for an Equation of the Mixed Elliptic-hyperbolic Type. *Dokl. Akad. Nauk, SSSR 101*, 793—796.

Keldysh, M. V.

(1) (1951) On Certain Classes of Elliptic Equations with Singularity on the Boundary of their Domain. *Dokl. Akad. Nauk, SSSR 77*, 2, 181—183.

Khalilov, Z. I.

(1) (1953) The Solution of a Problem for an Equation of Mixed Type by the Method of Nets. *Dokl. Akad. Nauk. Azrb.* SSSR *9*, 4, 189—194.

Khoi Chun'i

(1) (1958) The Dirichlet Problem for a Class of Second Order Linear Elliptic Equations which have Parabolic Singularity on the Boundary *Sci. Record*, *II*, 8, 244—249.

Kikvidze, Z. A.

(1) (1954) On a System of Partial Differential Equations of the Mixed Type. *(Soobshch. Akad. Nauk. Gruz. SSSR*, *15*, 6, 321—326.

Kudryavtsev, L. D.

(1) (1956) On the Solution of Elliptic Equations which are singular on the boundary by Variational Methods. *Dokl. Akad. Nauk, SSSR*, *108*, 1, 16—19.

Lavrent'ev, M. A. and Bitsadze, A. V.

(1) (1950) On the Problem of Equations of the Mixed Type. *Dokl. Akad. Nauk. SSSR 70*, 3, 373—376.

Ladyzhenskaya, O. A.

(1) (1954) Note on an Approximate Solution of the Lavrent'ev—Bitsadze Problem. *Usp. mat. nauk. 9*, 4 (62), 187—190.

Lèvy, E. E.

(1) (1905) Sulle Equazioni Lineari Totalmante Ellitiche alle Derivate Parzieli. *Rend. Circ. Mat. Palermo*, *24*, 275—317.

MIKHLIN, S. G.
(1) (1954) On the Theory of Singular Elliptic Equations. *Dokl. Akad. Nauk SSSR*, *94*, 3, 183—186.
MIRANDA, CARLO
(1) (1955) Equazioni alle Derivati Parziali di tipo Elliptico.
MORAWETZ, CATHLEEN, S.
(1) (1954) A Uniqueness Theorem in Frankl's Problem. *Commun. Pure Appl. Math.*, *7*, 4, 697—703.
(2) (1956) Note on a Maximum Principle and a Uniqueness Theorem for an Elliptic-hyperbolic Equation *Proc. Royal Soc.*, *232*, *A*, 141—144.
(3) (1957) Uniqueness for the Analogue of the Neumann Problem for Mixed Equations. *Michigan Math. J.*, *4*, 5—14.
(4) (1958) A Weak Solution for a System of Equations of Elliptic Hyperbolic Type. *Commun. Pure Appl. Math.*, *XI*, 3, 315—331.
MUSKHELISHVILI, N. I.
(1) (1946) Singular Integral Equations (Moscow—Leningrad, Gostekhizdat)
NIKOL'SKI, S. M.
(1) (1943) Linear Equations in Linear Normed Spaces. Izv. *Akad. Nauk SSSR*, *ser. mat.* *7*, 3, 147.
OLEINIK, O. A.
(1) (1952) Note on Elliptic Equations which are Singular on the Boundary. *Dokl. Akad. Nauk*, *SSSR* *87*, 6, 885—887.
(2) (1952) On Elliptic Equations of the Second Order *Usp. mat. nauk* *7*, 3, 106—107.
OU SING-MO and DING SHIA-SHI
(1) (1955) Sur l'Unicité du Problème de Tricomi de l'Equation de Chaplygin, *Sc. Rec.* *3*, 393—399.
OVSYANNIKOV, L. V.
(1) (1953) On the Tricomi Problem in a Class of Generalized Solutions of the Euler Darboux Equation. *Dokl. Akad. Nauk*, *SSSR* *81*, 3, 457—460.
PROTTER, M. H.
(1) (1954) The Cauchy Problem for Hyperbolic Second Order Equation. *Can. J. Math.* *6*, 4, 542—543.
(2) (1953) Uniqueness Theorems for the Tricomi Problem and (1955) I—II. *J. Rat. Mech. a Anal.* *2*, 107—114, *4*, 721—732.
(3) (1954) New Boundary Value Problems for the Wave Equation and Equations of the Mixed Type. *J. Rat. Mech. a Anal.*, *3*, 4, 435—446.
(4) (1954) An Existence Theorem for the Generalized Tricomi Problem. Duke Math. J. *21*, 1—7.
RUDNEV, G. V.
(1) (1951) On Certain Plane-parallel settled Gas Motions, Doctoral thesis, (Library of Math. Inst. Akad. Nauk, SSSR).
SEDOV, L. I.
(1) (1950) Problems in the Plane in Hydrodynamics and Aerodynamics. Moscow—Leningrad, Gostekhizdat.
SMIRNOV, M. M.
(1) (1955) On a Boundary Value Problem for an Equation of the Mixed Type. *Dokl. Akad. Nauk*, *SSSR* *104*, 5, 699—701.
SOMIGLIANA, C.
(1) (1894) Sui Sisteme Simmetrici di Equazioni a Derivate Parziali. *Ann. Mat. Pura ed apl.*, Ser. *II* 22, 143—156.

TERSENOV, S. A.
(1) (1957) On an Elliptic Equation which is Singular on the Boundary. *Dokl. Akad. Nauk, SSSR 115*, 4, 670—673.

TRICOMI, FRANCESCO
(1) (1923) Sulle Equazioni Lineari alle Derivate Parziali. di 2 Ordine di Tipo Misto. *Mem. Lincei*, Ser. *V*, *XIV*. fasc. *VII* : Translated into Russian (O lineinykh uravneniyakh smeshannogo tipa) Moscow—Leningrad, Gostekhizdat, 1947.
(2) (1928) Ulteriori Ricerche Sull'equazione $y\ \delta^2 z/\delta x^2 + \delta^2 z/\delta y^3 = 0$. *Rend. Circ. Mat. Palermo, LII*. [cf. supplement I to the Russian translation (1)].
(3) (1927) Ancora Sull'equazione $y\ \delta^2 z/\delta x^2 + \delta^2 z/\delta y^2 = 0$. *Rend. Acc. Lincei*. Ser. *VI*, [cf. supplement II to the Russian translation (1)].
(4) (1928) Sull'equazione $y\ \delta^2 z/\delta x^2 + \delta^2 z/\delta y^2 = 0$; *Atti Congr. Internaz., III*, Bologna [cf. supplement III to the Russian translation (1)].
(5) (1954) Lezioni Sulle Equazioni a Derivate Parziali, Torino,
(6) (1957) Equazioni a Derivate Parziali. Roma,
(7) (1954) Beispiel einer Strömung mit Durchgang durch die Schallgeschwindigkeit. *Monatsch. f. Math.*, *58*, 160—171.

VAN-GUAN-IN'.
(1) (1955) On the Uniqueness of the Tricomi Problem for Chaplygin Equation. *Acta, Math. Sinica*, *5*, 455—461.

WEINBERGER and HANS, F.
(1) (1954) Sur les Solutions Fortes du Problème de Tricomi. *C. R. Acad. Sci.* Paris, *238*, 1961—1962.

VEKUA, I. N.
(1) (1948) New Methods for the Solution of Elliptic Equations. Moscow, Gostekhizdat.
(2) (1953) Generalized Analytic Functions and their Applications. Moscow, Fizmatgiz.

VISHIK, M. I.
(1) (1951) On Strongly Elliptic Systems of Differential Equations. *Mat. Symposium 29*, 3, 515—676.
(2) (1955) On the First Boundary Value Problem for elliptic Differential Equations which are Singular on the Boundary of the Domain. *Dokl. Akad. Nauk, SSSR 93*, 1, 9—12.
(3) (1954) Boundary Value Problems for elliptic Equations which are Singular on the Boundary of the Domain. *Mat. Symposium : 35*, (77), 513—538.

VVEDENSKAYA, N. D.
(1) (1953) On a Boundary Value Problem for Equations of the Elliptic Type that are Singular on the Boundary of the Domain. *Dokl. Akad. Nauk SSSR 91*, 4, 711—714.

ZAREMBA, S.
(1) (1910) Sur un Problème Mixte Relatif à Equation de Laplace. *Bull. Cracovie*, Ser. A, 313—344. Translated into Russian *(Usp. mat. nauk*, 1946, *1*, 3—4 (13—14), 125—146.

SUBJECT INDEX

NAME INDEX